Before *Silent Spring*

Before
Silent Spring

Pesticides and Public Health
in Pre-DDT America

JAMES WHORTON

Princeton University Press PRINCETON, NEW JERSEY

Copyright © 1974 by Princeton University Press
Published by Princeton University Press
Princeton and London

ALL RIGHTS RESERVED

Library of Congress Cataloging in Publication Data
will be found on the last printed page of this book

Composed in Linotype Baskerville and printed
in the United States of America by
Princeton University Press
Princeton, New Jersey

for my MOTHER *and* FATHER,
for their example and encouragement

Preface

So MUCH LITERATURE dealing with the adverse effects of pesticides on human and animal life has appeared in the years since the 1962 publication of Rachel Carson's *Silent Spring* that the author of yet another work on the subject feels obligated to provide an apology. After *Silent Spring*, *Pesticides and the Living Landscape* (R. Rudd, Madison, 1964), *Zilveren Sluiers en Verborgen Gevaren* (C. Briejer, Leiden, 1967), and *Since Silent Spring* (F. Graham, Boston, 1970), there would seem to be little left to say about "the pesticide problem." Certainly those who have closely followed the recent pesticide debate will find little that is surprising in this book's examples of careless and uncontrolled application of dangerous insect poisons; of agricultural specialists making self-serving analyses of medical problems outside their areas of expertise; of those with medical expertise disputing one another's evaluations of environmental poisoning hazards; or of public health officials allowing their policy to be dictated more by political and economic considerations than by medical ones.

Yet if the lessons it contains are familiar, they are also worth repeating, if only to place in historical perspective our present dilemma of seemingly having to poison our food in order to protect it. Rachel Carson was fully aware that the abuses associated with DDT and kindred organic insecticides were largely the consequence of attitudes and practices formed during the past century's experience with inorganic insect poisons, but too many of the self-styled "ecologists" whom she roused to activity have viewed the problem through wholly contemporary glasses. Even knowledgeable and responsible writers on the subject have been guilty of historical simplification. A British expert on pesticides has recently proposed that the history of crop pest control might be divided into two periods—BC, Before Carson, and AC, After Carson—and suggested that not until these latter days did man realize "that pesticides were not the panacea heralding the millennium."[1] Another author has submitted that "The publication of *Silent Spring* . . . marked the end of closed debate in this field. Rachel Carson uncovered the hiding places of facts that should have been disclosed to the public long before; she broke the information barrier."[2]

In fact, the last century of the BC period witnessed a number of experts expressing feelings about pesticides that fell far short of chiliastic enthusiasm, and during the final decades of the era these misgivings were aired publicly. They did not, admittedly, excite quite the furor that followed *Silent Spring*, but then they were never presented at book-length by an author with the scientific and literary credentials of Miss Carson. Nor was the BC audience in as receptive a frame of mind as their grandchildren would be. The pre-World War II public was distracted from the insecticide hazard and many other problems by a deep economic depression, and Americans had not yet been conditioned to regard science and technology with a skeptical, even jaundiced, eye. Public awareness of the dangers of insecticides was thus not so great before Carson as after, but the awareness and the dangers existed nonetheless, and an understanding of how both developed is essential to a complete appreciation of the contemporary pesticide controversy.

The history of pesticides and public health is of interest not solely as a prelude to *Silent Spring*,[a] but also as an illustration of the difficulties of recognizing and controlling a type of public health problem that is coming to characterize the present age. Through all the centuries up to this one, the most serious epidemics have been those of microbic diseases, as the germs of plague, smallpox, yellow fever, cholera, tuberculosis, and a host of others have taken their turns at devastating Western society. The defeat of these enemies by the discoveries of late nineteenth century bacteriology is the greatest triumph, and the most familiar chapter, in the history of medicine. But just as the germs were being vanquished, the new hazard of chemical contaminants was emerging as an epidemic threat, and a threat less easily perceived than previous ones. An outbreak of bubonic plague, for example, impresses even the most casual observer as a serious menace to public health. Chemical pollutants, on the other hand, are generally present in too small amounts to produce immediate, acute illness. Rather, their insults to the body are of a chronic nature, taking months or years to manifest themselves, or perhaps never becoming obvious at all. Chronic poisoning may advance no further than a general reduction in vigor more readily attributable to the effects of aging or the stresses of modern living than to a specific contaminant of the victim's environment (particularly since that contaminant may no longer be detectable in the victim's body). Similar considerations apply to even the obvious maladies. The chemical

---

a The thrust of *Silent Spring* was actually somewhat broader than that of the pre-DDT pesticide agitation. As its title implied, Miss Carson's book was as concerned with the effects of insecticides on lower animals like birds as with their direct effects on man. Alarm over the inorganic insecticides used during the earlier period, however, was aroused almost entirely by the fear that people might be poisoned.

induction of cancer, for instance, is a process that may not complete itself until years after final exposure to the offending substance, in which case definite identification of the carcinogen can be nearly impossible.

Laboratory demonstrations of chemical etiology (as opposed to clinical ones) are not necessarily simpler or more conclusive. Since chronic illness may be so slow to develop, chronic toxicity studies ideally require that the experimental animal be regularly exposed to small amounts of the chemical in question through its entire lifetime. For the most commonly used animal, the white rat, the life expectancy is two years. It is also desirable to perform studies on additional species of animals, such as the dog, though for longer-lived animals the study must be abbreviated to only a fraction, perhaps one-fifth, of the lifetime. To insure the significance of the findings, furthermore, large numbers of animals of each tested species must be used, and similarly large control groups (animals living under identical conditions except for not being exposed to the tested chemical) have to be maintained. For every animal in each group, the investigator must determine such factors as rate of growth, changes in composition of blood and urine, reproductive ability through several generations, and pathological changes (found through gross and microscopic examinations of all tissues of the sacrificed animals).

Such studies are expensive, but not always of such clear urgency as to attract the necessary financial support. The question of urgency, moreover, is not determined simply by scientific considerations. The manufacturer of a new product may feel getting the product to market is more urgent than conducting time- and money-consuming searches for some possibly injurious effect of the product. Unless legally compelled, the manufacturer cannot be expected to conscientiously analyze the possible chronic effects of his food or other products before sending them to market; such legal compulsion did not exist during the period covered by this study. When industry declines to evaluate the safety of its products, the matter is left in the hands of either individual scientists, who have other research problems competing for their attention, or government health officials. The latter group has an interest in chronic toxicity determinations, but may lack the legal power or the financial resources to pursue them. Such situations will appear in the chapters to follow.

But when all the obstacles to toxicity studies are overcome, and a substance is tested and found to be injurious to animals, there will always be those who observe that rats and guinea pigs are not people, and that the contaminant is not therefore clearly a threat to public health. Their insistence that the final toxicological studies be made on human populations dramatizes the difficult choice being forced upon developed societies trying to maintain control of their chemical environment: should a chemical be permitted general circulation until its harmfulness becomes evident in clinically observable injury to people, or should it be prohibited from distribution because its effects on laboratory animals suggest it *might* be harmful to man? Is the "play it safe" philosophy a barrier to, or a governor of, technological progress? Even where the latter attitude prevails, and laboratory evidence of toxicity is accepted, there remain the questions of what quantity of a contaminant should be considered dangerous for man, and whether or not it occurs above the danger level in the environment.

Recognizing that a hazard exists is, of course, only half the solution of any public health problem, and quite often the easier half. Regulating the introduction of contaminants into the environment so as to reduce the concentration of each to an acceptable level, and keep it there, can be most frustrating. Chemical pollutants are generally the products of essential industrial processes, and sometimes the process is impossible without the pollutant. Even where there exists the technology to prevent the production or escape of the pollutant, there is rarely an eagerness to use it. The recent evasions of automobile manufacturers and smelter operators, among others, are so familiar as to make elaboration on this point superfluous. Pollution control programs have always to face stiff economic opposition, and the agricultural industry of half a century ago, it will be seen, could be as unyielding as any special interest group of later years.

A recent analyst of the contemporary pesticide debate has observed that "there is no such thing as an 'objective' book on an issue of social importance,"[3] and, at least with regard to the issue of pesticides, this generalization would appear valid. Objectivity has been at a premium in the pesticide controversy, and too often protagonists on both sides have yielded to simpler, emotional evaluations of their opponents. Conservationists have pictured industrialists as profit-driven environmental plunderers, while industry has scoffed at its critics as reincarnated Chicken Littles.

If we look at the problem historically, it is easier to divest oneself of the passions aroused by present "issues of social importance," and to see that blame is not to be carelessly laid on one side or the other. Chemical insecticides were developed to meet a pressing, legitimate need; they were developed at a time when the danger of epidemic chronic intoxication from environmental contaminants could not be fully appreciated. By the time this danger was appreciated, chemical pesticides were a *fait accompli*; predictably, agriculturalists, convinced that insecticides, as commonly used, were both necessary and safe, reacted strongly to charges that they were neither. Close inspection of the commotion that followed convinces one of the wisdom of a French physician of the supposedly more innocent days of 1912. "It is difficult to reconcile," he commented tersely, "the exigencies of public health with the desiderata of modern agriculture."[4] Just how difficult a reconciliation this could be is the subject of the following pages.

Intellectual debts are paradoxical: they are both the most difficult, and the most satisfying, to attempt to repay. My creditors for this work are many, and I can only hope they will be pleased with the use I have made of their knowledge and advice. Any failures to make full capital of their contributions are, of course, my own.

Professor Aaron J. Ihde of the University of Wisconsin introduced me to the history of chemistry, guided me through the initial exploration of the history of the pesticide problem, kindly criticized my manuscript, and encouraged me throughout; my obligation to him is deep and lasting. Professor James Harvey Young of Emory University has been just as encouraging and critical, and I deeply appreciate his aid. Thomas Dunlap is conducting, at the University of Wisconsin, a similar study of the DDT period, and has provided many useful suggestions of sources and interpretations, and criticisms of the manuscript's content and style. Judy Lowe of Seattle offered similar suggestions from a layman's viewpoint. Finally, Professor Charles W. Bodemer, of the University of Washington, has my gratitude for his help in securing free time and funding to make this book practicable.

Staff members of the National Archives in Washington, D.C., and of the National Records Center in Suitland, Maryland, were resourceful in locating and cooperative in providing the government records related to pesticide regulation. I am ever grateful to Mr. George Brite, who, as an economics specialist in the Congressional Research Service, helped me find needed volumes in the Library of Congress, and, as my uncle, has been an intellectual stimulus for as long as I can remember.

Ms. Sandra Johnson and Ms. Lynn Leonard know my gratitude to them for their invaluable secretarial aid. Special thanks are due Norma Baker, my exceptionally conscientious and cooperative typist. Mrs. Gail Filion and Miss R. Miriam Brokaw, editors at Princeton University Press, have been most interested and helpful in seeing this work through the press.

For their support for research costs, I am thankful to the National Science Foundation (Graduate Fellowship program), the Josiah Macy, Jr. Foundation (Program in the History of Medicine and Biological Sciences), and to the Graduate School and the department of History of Science of the University of Wisconsin. The typing of the manuscript was financed by the University of Washington.

For her patience, indeed endurance, at some stages of the project, for her careful proofreading and stylistic recommendations, for the complete sustenance she has provided from the beginning, I am above all grateful to my wife Sue.

Contents

Part One

Recognition

Which everywhere the farmer vexes,
And all philosophy perplexes.
Go where you will and search around,
Where'er they cultivate the ground,
In every land this curse prevails,
And all attempt against it fails.

—Prairie Farmer, March, 1851

The Insect Emergency 1

FARMERS HAVE ALWAYS LIVED at the mercy of the Nature they hope to exploit, determined to rule the land yet subject to the vagaries of soil and sun, wind, and rain, to sudden outbreaks of crop diseases, to unpredictable invasions of insects and other pests. The complaint of the prairie farmer above was provoked by the terrible potato blight of the 1840s, but it might have been directed with even greater accuracy at the plagues of the following decades, at the insect ravages that were to level not only potato fields, but to truly vex farmers everywhere.

Insect depredations were of course not uncommon before the mid-1800s. Visitations of locusts had been regular and devastating since Biblical days and, in recent years, American farmers had suffered heavy losses to the Hessian fly and the grain midge. Still, these damages had been only preparatory to the destruction reserved for modern times; earlier agriculture simply had not provided insects the opportunities for havoc soon to be offered.

The Agricultural Revolution, which so transformed the English countryside during the eighteenth century, was somewhat delayed in its arrival in America.[1] Statesman-farmer George Washington partially explained why in his

3

answer to the English farmer who "must entertain a contemptible opinion of our husbandry, or a horrid idea of our lands, when he shall be informed that not more than eight or ten bushels of wheat is the yield of an acre; but this low produce may be ascribed, and principally, too, to a cause . . . namely, that *the aim of the farmers in this country* . . . is not to make the most they can from the land, which is, or has been cheap, but the most of the labour, which is dear; the consequence of which has been, much ground has been scratched over and none cultivated or improved as it ought to have been; whereas a farmer in England, where land is dear and labour cheap, finds it his interest to improve and cultivate highly, that he may reap large crops from a small quantity of ground."[2]

The primitiveness of American agriculture in Washington's time was due not merely to the expense of labor but, more fundamentally, to the lack of a large market for farm products. In a society in which more than 90 percent of the population was composed of farm families, production for home consumption rather than commerce was the practical way to make a living. But between 1790 and 1860, city dwellers grew from 3.3 to 16.1 percent of the population,[3] and their demand for food could be satisfied only by a commercialization of agriculture. The necessary promotion of the farmer from provider to businessman was hastened, furthermore, by the astonishing system of transportation developed by mid-century. The turnpikes and canals opened during the 1820s through the 1840s stimulated trade between the eastern and western sections of the country, and the barge and steamboat were soon superseded by the railroad train. More than 30,000 miles of track stretched across the United States by 1860, and farmers as far west as the Iowa prairie had ready access to the markets of the eastern seaboard.

Social and technological changes thus encouraged the adoption of the revolutionized agriculture of England, with its scientific crop rotation schemes and conscientious appli-

4

cation of fertilizers. The price of day labor continuing high, however, profitable agriculture also demanded less time-consuming methods of sowing and reaping. From steel plows and seed drills, to steel-toothed cultivators and the Hussey and McCormick reapers, agricultural inventions by the thousands followed the call for labor-saving machinery (more than 10,000 U.S. patents had been granted for harvesting machines alone by 1880). It has been estimated that the farmer of 1860 who employed the latest equipment could grow and harvest his crops with only two-thirds the labor he would have used twenty years earlier.[4]

Unfortunately, as the critics of the machine age insisted so confidently with respect to the factory system, the new agriculture established conditions conducive to its own destruction. Mechanization fostered intensification and specialization in farming. Monocultural mass production appeared as the surest path to profits, and if few farmers could afford the huge "bonanza farms" of the northwestern wheat regions, many could at least imitate them on a modest scale. It was only through bitter experience that they learned that if large-scale cultivation of individual crops was the quickest way to pad the pocketbook, it was also a quick way to empty it. Extensive, unbroken fields canopied by the foliage of a single plant afforded ideal feeding conditions for the insect pests of that plant, and farmers whose livelihoods depended on the success of but one crop could be financially broken by a serious insect attack.

The favorable insect environment created by monoculture was further enhanced in America by westward expansion. The fulfillment of Manifest Destiny not only involved an enormous increase in the area of land under cultivation but, also, by the prerequisite clearing of forests in many areas, frequently destroyed predators of insects while forcing the insects themselves to turn to a domestic food supply. Even the marvelous new transportation system conspired with the insects, for the increase in interstate and international commerce too often allowed the transportation of

5

insects from one state or country to another in which there existed no natural predators of the pest. In this way, the European corn borer was introduced into America and, in unintended retribution, the Colorado potato beetle into Europe. In environments free of natural checks, these and numerous other displaced species were able to live and reproduce, and destroy, with virtual impunity.

It was thus that American farmers throughout the second half of the nineteenth century found themselves besieged by such unlikely sounding foes as the currant worm, the chinch bug, the codling moth, the cotton army worm, the Colorado potato beetle, the plum curculio, and, most fearsome of all, the Rocky Mountain locust or Western grasshopper. The last adversary launched attacks of such fury during the middle years of the 1870s that the governor of Missouri was compelled to proclaim a day of public prayer and fasting "for the interposition of Divine Providence to relieve the calamities caused by the devastation of the Rocky Mountain locust,"[5] and the federal government had to send emergency shipments of food, clothing, and seeds to hundreds of Mississippi Valley farmers. The reports of these farmers, reluctant witnesses of the grasshopper's ravages, repeatedly described such occurrences as the appearance of locusts in "large swarms like masses of clouds,"[6] swarms that "crackle beneath the feet of persons walking over the prairies,"[7] and that "often impeded the trains on the Western railroads . . . the insects passing over the track or basking thereon so numerously that the oil from their crushed bodies reduced the traction so as to actually stop the train, especially on an up-grade."[8] After allowing for the romantic exaggeration common to nineteenth-century pathos, one is still touched by the jeremiads of the locusts' victims:

"The farmer plows and plants. He cultivates in hope, watching his growing grain, in graceful, wave-like motion wafted to and fro by the warm summer winds. The green begins to golden; the harvest is at hand. Joy lightens his labor as the fruit of past toil is about to be realized. The day breaks with a smiling sun that sends his ripening rays through laden orchards and promising fields. Kine and stock of every sort are sleek with plenty, and all the earth seems glad. The day grows. Suddenly the sun's face is darkened, and clouds obscure the sky. The joy of the morn gives way to ominous fear. The day closes, and ravenous locust-swarms have fallen upon the land. The morrow comes, and ah! what a change it brings! The fertile land of promise and plenty has become a desolate . . . stretch of bare spindling stalks and stubs . . . and old Sol, even at his brightest, shines sadly through an atmosphere alive with myriads of glittering insects."[9]

A few years before, such eloquence would have been wasted as a cry in the wilderness, but by the 1870s the farmer's ancient lament was being enthusiastically answered by a new ally, the economic entomologist. Economic entomology was a fledgling science, an application of entomological knowledge of the structure and habits of insects to the economic problem of preventing or reducing insect depredations. It had been born during America's period of flagrant altruism, when every educated man felt duty-bound to be useful, and the first generation of economic entomologists embraced as their mission nothing less than the salvation of agriculture. They were in a sense goaded into their optimism, for the entomologist had never cut a very admirable figure before the public eye. In the century in which scientists generally were rising toward a status of popular heroes, entomologists continued to be looked upon as bearded, spectacle-wearing eccentrics who wasted their days darting about the fields waving butterfly nets. People chuckled in agreement with Oliver Wendell Holmes when he defined the entomologist as a man "who gives insects long names and short lives, a place in science and a pin through the body."[10]

Entomologists were aware of their pitiful image. "The idea of the trifling nature of his pursuit," one analyzed, "is so strongly associated with that of the diminutive size of its

objects, that an *entomologist* is synonymous with everything futile and childish . . . how can he look for sympathy
in a pursuit unknown to the world, except as indicative of
littleness of mind."[11] The economic application of his science was too good an opportunity for self-vindication to be
missed, and as insect damages rose to unprecedented levels,
entomologists stepped forward to offer their services. At the
head of the line was Thaddeus William Harris, a man eventually to be recognized as "father of economic entomology
in America." Dr. Harris was a practicing physician and,
from 1831 until his death in 1856, Librarian of Harvard
College. Entomology was only his avocation, but he pur-
sued it with such ardor and competence that when the
Massachusetts Zoological and Botanical Commission desired a survey of the state's harmful insects, Harris was commissioned to do the work. His *Report on Insects Injurious
to Vegetation* was first published in 1841 and subsequently
reissued in several editions as a *Treatise* of somewhat longer
title and broader scope. Harris' *Treatise* was an encyclo-
pedic exposition of all the destructive insects a New England farmer was likely to encounter, complete with illus-
trations and suggestions for combatting each pest.

If Harris brought forth economic entomology in America, the labors of others were essential in guiding the disci-
pline to maturity. Dr. Asa Fitch, another professional
physician-amateur entomologist, became the first official
state entomologist when he was appointed to that position
by the New York legislature in 1854. Over the course of the
next sixteen years, Fitch issued fourteen reports on the life
cycles and habits of New York insects, reports that were
most useful in helping farmers identify and destroy some
of their worst enemies, and that one state senator estimated
had saved his constituents fifty thousand dollars annually.[12]
Such praise could only encourage other states to follow New
York's lead, and State Entomologist was a common title by
the end of the century. Most noteworthy among the early
appointees to the position were Benjamin D. Walsh of Illi-

nois and Charles Valentine Riley of Missouri, both of
whom assumed their duties in 1868. Walsh died the following year (not without having made his impress on the
developing science), but Riley lived on for nearly three
decades and filled his years with invaluable studies and
writings. Riley looked "much more like an Italian artist
than like an American economic entomologist,"[13] but be-
lied his appearance by becoming the man who, above anyone else, "sold the importance of economic entomology to
the public."[14]

Consideration of the early development of economic
entomology would be incomplete without mention of
Townend Glover, the first entomologist to receive a federal
appointment. The mid-nineteenth century witnessed a rising governmental concern for the welfare of agriculture,
first prominent in the 1839 congressional appropriation of
$1,000 for the U.S. Patent Office to collect agricultural sta-
tistics, investigate methods of promoting agriculture, and
distribute seeds to farmers.[a] The Commissioner of Patents
subsequently established an agricultural division within his
office, and staffed it with a chemist, a botanist, and an entomologist, Glover. Hired in 1854, Glover served for five
years, resigned to teach for three, then reassumed his position as government entomologist when Congress created an
independent Bureau of Agriculture in 1862. Glover stayed
with the Bureau (made a Department in 1876) until failing
health forced his retirement in 1878. During these years,
Glover traveled about the country collecting information
about the most injurious insects of each region, passed this

_____

[a] A. Hunter Dupree, *Science in the Federal Government*, Cambridge,
Mass., 1957, pp. 110-114, 149-183, offers an excellent survey of the
growth of federal support of agriculture. He comments further that
the early sponsorship of agriculture, culminating in 1862 with the
establishment of a Bureau of Agriculture and the passage of the Morrill Act to provide for land-grant agricultural colleges, was the major
turning point in government sponsorship of scientific research and
began the "era of bureau-building" (p. 151) in the federal government.

information on to farmers through the pages of the Bureau's *Annual Reports,* and replied to innumerable queries from puzzled agriculturalists. His contributions were even more remarkable in view of the non-entomological duties with which he was saddled by his superiors. Questions concerning such diverse matters as fruits, textiles, and birds were all channeled to Glover, and while he derived much satisfaction from the avian museum he developed, Glover and his fellow entomologists were much annoyed by the government's less-than-enthusiastic support for their science.[15]

Entomological pique was less important for winning increased federal support than was the intervention of the Rocky Mountain locust. The severe damages inflicted by that pest finally moved Congress to fund a United States Entomological Commission, a three-man board chaired by Riley and charged with the investigation of the locust, as well as other destructive insects. The Commission began its work in 1876 and produced two thick volumes on the locust, plus one each on cotton insects and forest insects, before metamorphosing into the Division of Entomology and then the Bureau of Entomology, in the Department of Agriculture. The most meaningful yardstick for measuring the rate of growth of federal support for economic entomology is the record of fiscal appropriations for entomological work. The amounts allotted fluctuated considerably from year to year, but after 1881 never dipped as low as the eighteen thousand dollars appropriated to establish the Entomological Commission. Fifty thousand dollars was surpassed for the first time in 1889, and two hundred thousand exceeded in 1907.[16]

The federal government encouraged the progress of economic entomology in less direct ways. The well-known Morrill Act of 1862 presented the individual states with public lands to provide income to finance agricultural schools and engineering colleges. Although a few agricultural schools had existed prior to the act, the newly formed land-grant colleges furnished a far broader base for the teaching of the agricultural sciences, including economic entomology. Of still greater import was the passage in 1887 of the Hatch Act for the establishment of state agricultural experiment stations in conjunction with the land-grant colleges. The need for research facilities at these colleges had led several states to support experiment stations prior to the Hatch Act, but the provision of federal funds dramatically increased the number of these stations and gave new impetus to entomological research. Whereas only three states appear to have given continued support to the position of state entomologist before the Hatch Act took effect, within five years after the founding of the experiment stations, forty-two states and territories could boast of official entomologists.[17] The importance of the research done by these scientists for the development of more effective insecticides can hardly be over-emphasized.[b]

During the pre-Hatch period, though, the contributions of entomology to agriculture fell some distance short of entomologists' expectations. Without adequate research facilities, entomologists had to rely on cooperative farmers for the testing of insect control procedures. As Glover reminded his readers, "the naturalist, studying the habits and instincts of injurious insects, may suggest remedies, but it remains to the farmer to aid him and make known the results."[18] The success of this partnership between scientist and husbandmen depended, of course, on two variables: the ability of entomologists to devise effective "remedies," and the willingness of farmers to test these and report their findings. Neither party, sad to say, wholly lived up to its half of the bargain.

Farmers proved to be frustratingly obstinate, too many

[b] There was, it should be noted, close cooperation between the federal Department of Agriculture and the many state experiment stations. The Department's Office of Experiment Stations was set up to furnish assistance and advice to the stations, and even to indicate areas most in need of research.

preferring their traditional ignorance to entomological enlightenment. The country's first journal of economic entomology, *The Practical Entomologist*, suggested in its initial number in 1865 that only by mastering the journal's information on "the Natural History of the Different Species of Insects"[19] could farmers hope to overcome their enemies. The message was repeated with increasing urgency over the next two years, up to the date the journal folded for lack of financial support. Undismayed by this first failure, and unwilling to believe that the agricultural community could continue to overlook the value of entomological advice, Walsh and Riley pooled their talents to produce the *American Entomologist* in 1868. It lasted exactly as long as its predecessor, and another decade was to pass before any journal devoted to economic entomology was to survive beyond infancy.

Before the farmers are branded refractory, however, the entomological advice they were ignoring should be examined. On the positive side, the suggestions for insect control were nothing if not abundant. The number of proposed remedies possibly exceeded the number of injurious insect species, but all could be assigned to one of three categories: biological, physical, and chemical methods of control. Biological measures, the reliance on other animals to destroy insect pests, comprised the smallest class. Observations that injurious insects increase in numbers when their natural predators are destroyed, or when the insects themselves are transported to an area free of predators, were commonplace in nineteenth-century entomological writings. The corollary that insect invasions could therefore be defeated by the introduction of the proper predators among the invaders was also suggested, and even occasionally applied to practice. Glover had a special interest in preserving birds that feed on insects, and worked such long hours organizing his bird museum primarily to demonstrate the "relations existing between the farmer's insect enemies and his feathered friends."[20] Glover and his fellow entomologists also stressed the part played by predatory insects in holding pest populations under control, and noted that European gardeners had for years stocked their plots with ladybirds (Coccinellidae) to control aphid infestations. A few similar biological controls were tested by American entomologists during the second half of the century with some success, but there was no clear-cut victory over a pest until the final decade. The victim in this case was the cottony-cushion scale, a native of Australia which had been accidentally brought to California in the early 1870s. The scale adapted to the California environment so well that it had brought the state's citrus industry to the brink of ruin by the time C. V. Riley came to the rescue. As chief of the federal Bureau of Entomology, Riley was most concerned about the ravages of the scale, and finally was able to wangle funds to send a colleague to Australia to search for predators of the scale. The envoy chosen for this mission was Albert Koebele, a young member of the Bureau's staff who had distinguished himself in field work and was familiar with the scale. Landing in Sydney in the fall of 1888, Koebele required but a few weeks to discover the vedalia beetle (Novius cardinalis), an insect that makes all its meals on the eggs and larvae of the cottony-cushion scale. Koebele sent several shipments of the vedalia back to California, all arrived safely, the beetles thrived, and by 1890 the cottony-cushion scale was a menace of the past.

The ease with which the scale was subdued was spectacular, and most atypical. A few other insects have since been brought under control by biological methods, but these victories have been difficult, accomplished only by painstaking study in both the laboratory and the field, and aided by not a little luck. During the years of the nineteenth century when the insect menace became an emergency, there was rarely sufficient knowledge or time to develop adequate biological control systems, and other types of remedies had to be employed in nearly all situations.

Farmers had always placed much more reliance on artificial than on natural controls anyway. A perusal of the entomological literature of the period quickly discloses that the large majority of insecticide techniques were rather crude physical measures. The *Treatise* of T. W. Harris remained the standard reference for economic entomologists into the 1870s, but even Harris' recommendations seem impractical for the large farm. If attacked by caterpillars, Harris advised, pay children to collect them by the quart; blister beetles can be removed by shaking them off foliage into deep pans; and to control wire worms Harris counseled his readers to sprinkle their fields with slices of potato or turnip and, on the following day, to collect and destroy the insects that had come to feed on the bait.[21]

Harris was not the farmer's only source of information. Other entomologists had recommendations of their own, but, if different, these were rarely less toilsome or time-consuming. Glover thought that syringing plants with whale-oil soap and then trampling underfoot the insects that fell from the slippery leaves was a good method. For the frugal, the procedure of placing coops of chickens beneath fruit trees and shaking the insects off the trees and, hopefully, into the coops, saved not only the expense of whale-oil soap, but of chicken feed as well.[22] Farmers, finally, did not rely entirely on entomologically approved techniques, but gave free rein to their own imaginations and generously shared their ideas with others through the medium of agricultural and entomological journals. A southern Colonel Hardee even drew upon his military experiences in suggesting that insects might be destroyed by concussion if enough gunpowder were exploded near the fields.[23] When his proposed method was tested against the Rocky Mountain locust, however, it proved singularly ineffective, and almost before the smoke had cleared was abandoned in favor of tried-and-true measures such as clubbing the locusts to death.

It should not be concluded that primitive hand-to-hand combat was the agriculturalists' last resort, for crude physical assaults had been reinforced for centuries by various forms of chemical warfare. Several insecticidal chemicals were recommended by the agricultural writers of antiquity, and their list was steadily lengthened through the ages. By the eighteenth century, there had accumulated a veritable "pharmacopoeia" of insect remedies that leaned quite heavily on herbal and animal preparations similar to those which dominated the official lists of drugs for human illness. Thomas Moufet, who also spelled his name Moffat and Muffet and is suspected of being the father of the little miss frightened by a spider, published the first English work on entomology in 1658, and included in *The Theatre of Insects* a "large field of remedies" provided by "most bountiful Nature," such as fern root, penny royal, rue, and "above all, the dregs of Mares-pisse."[24] These poisons were not restricted to field use, it might be added, but were deemed of equal efficacy in dealing with domestic pests. The Elizabethan agriculturalist Thomas Tusser advised against the then ubiquitous flea that:

When wormwood hath seed, get a handful of twaine
To save against March to make flea refraine;
Where chamber is swept and wormwood is strown,
No flea for his life dare abide to be known.[25]

Many of these presumed remedies were doubtless quite fanciful, but a few were actually effective and passed the test of time. By the middle of the nineteenth century, the most popular of the chemical insecticides were hellebore, quassia, lime, and tobacco, all applied as aqueous decoctions.

It was at about this point that a new substance of promise became known to European and American agriculturalists, one with an extraordinarily romantic history for an insecticide.[26] The pyrethrum flower, a member of the chrysanthemum family, had been cultivated for years in the Caucasus, where it was dried, powdered, and sold to neighboring peoples to the south as a powerful insecticide. The price of pyrethrum was quite high, not because of any special expense associated with its preparation, but simply because the Caucasian growers so carefully protected their monopoly on the plant. The very origin of the powder was successfully kept secret until the beginning of the nineteenth century, when an Armenian merchant named Iumtikov learned that the insecticide was made from pyrethrum flowers. Obtaining seeds of the plant and successfully cultivating them involved further difficulties, but by 1828 Iumtikov's son was manufacturing pyrethrum powder on a large scale. It was several years more before the insecticide was marketed outside the Middle East, and the 1850s before this "Persian Insect Powder" was imported into the United States. In keeping with its exotic name, the new insecticide was much too expensive to be used for anything except household pests, but its success in this limited sphere was so impressive as to recommend an attempt at domestic production. Thus the Department of Agriculture, in the spring of 1881, distributed pyrethrum seeds and instructions on their cultivation to cooperating farmers in all sections of the country. Whether the seeds were bad, or the Department's instruction faulty, or the farmers insufficiently diligent, or the American soil and climate too alien, the trial program failed to realize its founders' hopes. Many seeds never vegetated; of those which did, the resultant plants were often undersized, and even the healthy flowers were sometimes less lethal than anticipated. "A slug which ate my plants," a Norfolk, Virginia pyrethrum tester complained to the Department of Agriculture,[27] "was not injured by the meal." Only in California did pyrethrum prosper. On the Buhach Plantation near Stockton, a moderate level of pyrethrum production was achieved and, by the mid-1880s, "Buhach" had become a popular insecticide. Like its Persian predecessor, unfortunately, Buhach remained too expensive for any but limited use, and entomologists' predictions that pyrethrum would be "The insecticide of the future"[28] came to naught.

In the second half of the nineteenth century, the future of insect control lay not with organic substances like pyrethrum, but with inorganic poisons, and in particular with arsenic. Compounds of arsenic had been employed in small-scale battles against insects since at least the beginning of the Christian era, and were still being recommended, on a small scale, in the mid-1800s. Harris cautiously advised using mashed potatoes laced with arsenic to lure household insects to their doom, while entomologists with differing tastes suggested grated carrots in place of the mashed potatoes.[29] Mixtures of red lead and molasses and of cobalt and water were also used at times, and at least one entomologist, Glover, appreciated the possibility of expanding these specific poisoning techniques into a general program of insect control through chemicals. As early as 1855, the federal entomologist proposed that "the thing to be chiefly desired now is, to find out the favorite food of the particular kind of insect to be destroyed; then to discover and use some efficient poison for the accomplishment of the purpose."[30] Glover's vision was apparently too novel to be immediately shared by his colleagues. A decade later, the *Practical Entomologist* launched its brief career by proclaiming that "if the work of destroying insects is to be accomplished satisfactorily, we feel confident that it will have to be the result of no chemical preparations, but of simple means, directed by a knowledge of the history and habits of the depredators."[31]

Despite the entomological profession's confidence in the tried and true, changing circumstances were forcing a departure from the traditional "simple means" of insect destruction, one that went even beyond the use of poisoned bait recommended by Glover. These new circumstances were the work of the Colorado potato beetle, an insect native to the Rocky Mountains which had thrived for centuries on the leaves of wild plants. When the white man moved into Colorado, the beetle deserted its traditional diet in favor of the more palatable foliage of the Irish potato brought by the settlers. Colorado potato patches were but a springboard to the more developed agricultural areas to the east, and by the late 1850s the beetle was off and moving on an inexorable march to the sea characterized by its own version of scorched earth policy. The *Practical Entomologist* warned subscribers in 1865 that the Colorado beetle had already advanced into Illinois and Wisconsin, while along the way "committing the most destructive ravages on the potato crop . . . so as to threaten the loss of the entire crop on many farms."[32] Understandably, keeping track of the beetle became a rural preoccupation; reports on its path and rate of progress were soon a regular feature in agricultural magazines, as were answers to queries about its appearance and habits and means of destroying it. Anxious farmers even mailed suspected beetles to their preferred journalists for identification, a potentially dangerous practice that agricultural editors discouraged. In 1868, the *American Agriculturalist* pleaded with its subscribers not to "send us any more specimens of this insect. We know it by sight thoroughly. A few days ago we received a package which contained a crushed box with some hundreds of these fellows all alive. Had the paper broken, the insects would have found their way out of the mail bag, and their eastern march would have been more rapid than it is now."[33] One year later, the same editors announced: "That which we feared has been done. The Colorado Potato Beetle has been scattered along our Eastern States. A friend in Paulding, Ohio, sent specimens in a thin pasteboard box which reached us in a smashed condition, with one remaining larva to show what it had contained. . . . The beetles have escaped, and we may look for them anywhere in the East."[34]

The Colorado potato beetle was indeed in the East, as well as all over the Mid-West, and farmers in both regions were in frantic search of means to defend themselves. Just how frantic can be appreciated by the discussion in a leading agricultural journal, of reports that "come to us from Western readers of sharpers who go about the country selling packages highly recommended as a 'simple, sure, cure for Potato Beetles.' On the outside the parcels are labeled: 'Don't open to expose to the air until ready to use,' and

'Directions for use inside.' After the swindlers are at a safe distance, the purchaser being ready to apply the 'sure cure,' finds, on opening the parcel, two blocks of wood with the 'direction': 'Put the beetles on one block and mash them with the other.' "[c]

The gullibility of some rustics was fortunately countered by the mechanical ingenuity of others, and more sophisticated insect-killing machines than wood blocks were soon made available. Some were true mechanical marvels, horse-drawn vehicles capable of dislodging beetles from potato plants and then crushing the insects under heavy rollers. The editor of the *Practical Entomologist* was so excited by these engines of destruction that he could scarcely contain his pride and joy: "The world certainly does move. Who would have believed fifty years ago that, in the year 1867, we should be slaying bugs by Horsepower?"[35] The entomologist's astonishment at the progress of civilization would have been even more wide-eyed, however, had he realized that in the very year of his boasting, in response to the same potato beetle, desperate farmers were at last turning to the approach to insect control that would shortly make even "horsepower" obsolete.

---

[c] *American Agriculturalist, 41,* 100 (1882). Similar reports continue to come out of Western states. The *Seattle Times* on September 17, 1970 carried the story of a housewife who sent a Dallas mail-order house $2.98 for a "sure-fire roach killer." "A few days later," the article relates, "the product arrived: Two blocks of wood." The president of the local Better Business Bureau responded to the housewife's complaint with the observation that "Old swindles never die." He elaborated by noting that the " 'Sure-fire bug killer' swindle was around more than twenty years ago when the device generally arrived with instructions reading: 'Catch the bug and place it on the wood, then smash it.' "

Horse-drawn beetle crushers were as impractical as they were impressive. They excited the American reverence for gadgetry, but they cost money and required time and care to use. What was demanded by the average farmer was a method of control that was inexpensive and that could be quickly applied to the whole farm. Cheapness and ease of application, in addition to effectiveness, were the prerequisites for a popular insecticide, and the right combination was finally found in Paris green, a then common arsenic-containing pigment. Chemically speaking, Paris green (alias Schweinfurth green, alias Vienna green) is a copper acetoarsenite (approximately $3Cu(AsO_2)_2 \cdot Cu(C_2H_3O_2)_2$). It and Scheele's green (copper arsenite, $CuHAsO_3$) were used to color most of the green paints in the nineteenth century. (While we are immersed in chemical nomenclature, we should note that the term "arsenic" has been used rather casually by chemical writers. Strictly speaking, the term denotes a gray, metallic-looking element, but it has more often been used to refer to the most common arsenical compound, arsenic trioxide [white arsenic, As_2O_3], and it is this latter meaning that should be understood whenever the word arsenic appears in this narrative.)

The particulars of the first use of Paris green against insects are unfortunately beyond the historian's reach. The application of arsenic to his potato plants was an act of desperation taken by some unknown farmer (or farmers) who spread word of the discovery by mouth rather than pen. There is a rumor that the insecticidal properties of the pigment were discovered fortuitously by a farmer who, having just painted his window shutters green, disposed of the remaining paint by throwing it over his beetle-infested potato plants.[36] Whatever the circumstances, the indications are that Paris green began its career as an insecticide in the summer of 1867.[37] At first slow to catch on, by the end of the following season it had risen far enough in the farmers' regard to excite journalistic warnings of the dangers of mixing arsenic with agriculture: "The following is going the rounds of the press. 'Sure death to Potato Bugs: Take 1 lb. Paris green, 2 lbs. pulverized lime. Mix together, and sprinkle the vines.' We consider this unsafe, as there is no intimation of the fact, not generally known, that Paris green is a compound of arsenic and copper, and a deadly poison. Such things should never be recommended without a full statement of their properties, so that one may know with what he is dealing. The poison would be very likely to kill the potato bugs, but how about the vines."[38]

Despite such explicit warnings of danger to his crops, and implicit ones of danger to himself, the American farmer turned steadily toward Paris green, and his journalistic advisers followed. Less than three years after issuing the warning quoted above, the *American Agriculturalist* was enthusiastically endorsing Paris green as the best defense against the Colorado beetle.[39]

What works against one bug might be expected to work against another, and Paris green's triumph over the potato beetle recommended its use to combat other pests. Soon melons, squash, cabbage, a few other vegetables, most fruits, and cotton were being regularly given protective coats of Paris green, while improvements in the methods of application offered still further encouragement to adopt arsenical insecticides. Originally, Paris green had been applied in dry form, diluted with larger quantities of lime or ashes or flour, and dusted over the plants. The first five years' experience with the insecticide, though, revealed that application in the wet form, as a water slurry, was easier and safer. The sparingly soluble pigment was periodically stirred to keep it well-mixed, and spread over foliage with brooms or brushes, or even handfuls of grass. These were effective applicators, but primitive, and they were shortly replaced by more efficient syringes and spray pumps. Arsenic could now be broadcast over the land. It had become the ally of first resort whenever death must be dealt to any pest.[40] At the close of its first decade as an insecticide, Paris green was selling at the rate of more than five hundred tons a year in the New York City market alone.[41]

By this time, also, Paris green was no longer alone on the market, but was being rivalled by several other cosmopolitan-sounding compounds, products such as Paris purple, English purple poison, and, the green pigment's chief competitor, London purple. The London poison, a by-product of the aniline dye industry, was composed largely of calcium arsenite and, because toxic and presumably useless, had simply been dumped at sea for years. Once arsenic became *de rigeur* in rural circles, however, the London dye firm of Hemingway and Co. mailed packages of its poisonous waste-product to several American agriculturalists for testing as an insecticide. Dr. C. E. Bessey, botanist at Iowa State Agricultural College, was the first to experiment with the London sample, trying it against the Colorado potato beetle in 1878. He and his coworker, J. L. Budd, found that "a single application placed every one of the pests on their backs over the ground, either dead or in a dying condition, in less than six hours."[42] Only Paris green could match potency like that, and the older insecticide lacked certain advantages of the London poison. First of all, London purple (the name was suggested by Bessey) was more soluble than Paris green and did not have to be stirred as often during wet application. It adhered better to plant foliage, and thus did not have to be applied so frequently. Its more conspicuous color made its presence on foliage readily detectable, thereby guarding against accidents and making it easier for the farmer to know when another application was necessary. Finally, London purple was considerably cheaper than Paris green and, all things considered, it is not surprising that for a period London purple almost completely supplanted Paris green as the insecticide of choice. For nearly two decades, Hemingway dumped all of the "purple" it manufactured onto the American agricultural market instead of into the Atlantic, but eventually London purple fell from grace as quickly as it had ascended. Unfortunate experience revealed that the insecticide's greater solubility meant not only less troublesome application but also less healthy plants. Dissolved arsenic is readily absorbed by plant structures, and just as readily poisons them. With so much arsenic in solution, London purple sprays as often as not killed plant along with pest, and although efforts were made to produce milder versions of the spray, the hazard of foliage damage led to London purple's near total decline by 1900.

This decline did not leave the field uncontested to Paris green, for shortly before 1900 the most effective arsenical insecticide of all had been discovered, during the campaign against the gypsy moth. A leaf-eating insect native to Europe, the moth was first brought to the New World by Leopold Trouvelot, a French-born Harvard astronomer with a side-interest in silkworm breeding. Some experiments dealing with the latter subject seemed to Trouvelot to require gypsy moths, and so in 1869 he imported a number of gypsy moth eggs and cared for them until the insects reached maturity. The moths wasted little time asserting their independence, and soon escaped from Trouvelot's home in Medford, Massachusetts. Twenty years later, in 1889, their descendants, in the form of caterpillars, returned to Medford, in numbers that "were so enormous that the trees were completely stripped of their leaves, the crawling caterpillars covered the sidewalks, the trunks of the shade trees, the fences and the sides of the houses, entering the houses and getting into the food and into the beds. They were killed in countless numbers by the inhabitants who swept them up into piles, poured kerosene over them and set them on fire. Thousands upon thousands were crushed under the feet of pedestrians, and a pungent and filthy stench arose from their decaying bodies. The numbers were so great that in the still, summer nights the sound of their feeding could plainly be heard, while the pattering of the excremental pellets on the ground sounded like rain. . . .

Persons walking along the streets would become covered with caterpillars spinning down from the trees."[43]

The nightmare was not confined to Medford. Since its escape in 1869, the gypsy moth had gradually and insidiously infiltrated all of New England, and by 1890 so many thousands of valuable fruit, shade, and especially forest trees had been destroyed that the Massachusetts legislature authorized the governor to appoint a commission to attempt to exterminate the insect. Hopes for extermination were naturally placed in Paris green, but for once the arsenical had met its match. The caterpillars, it was learned, were exceptionally resistant to arsenical poisoning, and the quantities of Paris green required to kill them were so great as to cause nearly as much injury to foliage as that produced by the insects. An escape from the dilemma was at last provided by F. C. Moulton, a Gypsy Moth Commission chemist, who introduced lead arsenate (approximately $PbHAsO_4$) as an insecticide in 1892. Though more expensive than either Paris green or London purple, the lead compound was also much gentler to foliage, and its success in halting the advance of the gypsy moth won it employment in battles against other insect pests. By the early 1900s, lead arsenate had become the most popular of all insecticides, and was to retain this position until the post-World War II release of DDT for public consumption. The only serious rival to lead arsenate during the first half of the twentieth century was calcium arsenate, introduced in 1906 for use against cotton pests in the South.

The road to popularity was not smooth for any of the early insecticides. Arsenicals had detractors, and more than a few concerned farmers sympathized with the contributor to *Garden and Forest* who worried that "hundreds of tons of a most virulent mineral poison in the hands of hundreds of thousands of people, to be freely used in fields, orchards and gardens all over the continent, will incur what in the aggregate must be a danger worthy of serious thought."[44] In fact, much serious thought was given to the danger associated with the application of arsenic to crops, but while several distinct hazards were recognized, none was reckoned so serious as to proscribe the use of arsenicals.

The risk of arsenical damage to foliage has been noticed. Withered potato vines were an accompaniment to Paris green from the beginning, and though this damage could be minimized by dilution of the insecticide with water or an inert solid, foliage burn remained a difficulty until the appearance of lead arsenate. A second problem recognized early was the danger of arsenical poisoning of the soil. Paris green had not been in use three seasons when Townend Glover began experimental inquiries into the ability of plants to grow in arsenical soil. His preliminary findings, reported in 1870, indicated that arsenic in the soil retarded plant growth,[45] but, being an entomologist, Glover was hesitant to assume authority on chemical questions and left the definitive study of this matter to a colleague in Washington, the chemist William McMurtrie. McMurtrie's more thorough investigations were not completed until 1875, but his findings made the wait excusable.[46] The chemist reported that he had grown a number of pea plants in individual flower pots, each pot containing 91.5 cubic inches of soil plus varying amounts of Paris green. Observing that plant growth was significantly inhibited only after the amount of arsenic in the pot exceeded 500 milligrams, McMurtrie extrapolated from pot to field to conclude that Paris green could be applied to the extent of 906.4 pounds per acre (!) without damaging the soil. Further assuming that much of the arsenic from one year's application would be removed by drainage before the next season, McMurtrie argued that the limit of about 900 pounds per acre would never be approached in practice, and that the use of arsenical insecticides therefore posed no threat to agriculture. With these assurances, the question of the danger of arsenic to soil was dropped, only to bounce back into consideration, as will be seen, three decades later.

In the meantime, farmers had other worries. Arsenicals,

properly used, might be safe for plants and the soil, but what of their effects on livestock? The rise in favor of Paris green had been paralleled by an increasing number of letters to agricultural magazines complaining of farm animals being sickened or killed by the new insecticide. Even after a decade of use of Paris green, it could be reported that " 'Familiarity has bred contempt' for this dangerous substance, and there have been more horses, cows, pigs and fowls lost by poisoning than ever before."[47] Of the cases cited as illustration of the danger of arsenic, though, all the deaths had been due to gross negligence on the part of the animals' owners. A piece of paper in which Paris green had been wrapped, for example, had been carelessly thrown away and had blown into a pasture where it was sampled by a curious calf. But if mishaps such as this were easily preventable, there was another potential source of poisoning not so readily avoided. On many farms, it was necessary to graze livestock in the orchard, and as arsenical spraying of fruit became common, so did farmers' fears that their livestock might be poisoned by orchard grass.

A. J. Cook, entomologist at Michigan Agricultural College and the most active propagandist for arsenical insecticides, laid these fears to rest. In 1889 Cook reported that he had sprayed one of his fruit trees with double-strength London purple (one pound of insecticide to one hundred gallons of water) and allowed the tree to drip dry onto a large sheet of paper spread around its base.[48] Robert Kedzie, a physician-chemist and a faculty colleague of Cook's, analyzed the paper and found a quantity of arsenic too small, in Cook's opinion, to harm either livestock or pets. To lend his conclusion *éclat*, Cook then applied London purple to another tree, allowed it to drip onto the grass, and cut the grass and fed it to his own horse. As presumably expected, the animal survived the meal in good health, as did a sheep on whom the experiment was repeated.

Horses and sheep were unfortunately not the only economically important animals to be exposed to arsenical sprays. Apiculture was a thriving business when the new insecticides came into use, and was the first area in which truly serious injury by arsenicals became apparent. In retrospect, it seems obvious that honey bees would be vulnerable to insecticides, and that caution in the use of arsenic should be the rule in bee-keeping areas. To those exuberant farmers first getting hold of the new insect poisons, though, the danger was obscured by ignorance and wishful thinking. The role of bees in fruit pollination seems not to have been widely appreciated until the last quarter of the nineteenth century, and, prior to that time, the bee was as often considered an enemy as an aid. Orchardists argued that bees injured fruit by sucking out vital juices and leaving it to wither. Beekeepers disputed the charge and demanded an end to the practice of setting out dishes of poisoned honey in orchards and vineyards.[d] Entomologist Cook, it should be noted, was especially active in educating fruit growers to the usefulness of bees and thus discouraging their intentional poisoning.[49]

Cook remained better known, however, for his advocacy of arsenical insecticides, and was so persuasive in this area that bees continued to die in droves long after the last cup of poisoned honey had been removed. In their eagerness to protect fruit completely from harmful insects, and particularly from the codling moth, many farmers sprayed their trees too early in the season, while the trees were yet in bloom and being visited by bees (and before, incidentally, the codling moth had laid its eggs, which it does only after the fruit blossoms fall). In some areas, more bees were being destroyed than moths, and the pages of the *American*

d The editor of the *American Bee Journal*, 23, 803 (1887) warned that poisoned bees might survive long enough to return to the hive and poison their own honey. "Would not the instigator of this plot be held liable for any damage done to humanity as a result of eating this poisoned honey?" A less restrained contributor to the next year's volume 24, 40 (1888) proposed feeding the "villain . . . on the same stuff that he recommends for the destruction of the honey bee."

Bee Journal during the 1880s overflowed with angry letters from apiculturalists ruined by a fruit-growing neighbor's over-zealous application of Paris green.

Orchardists refused to accept these claims. They were inclined to agree with the New York entomologist J. A. Lintner that the arsenical spray never penetrated to the nectar in the blossoms, and could therefore not be ingested by bees.[50] An Indiana farmer even announced that since he had been using Paris green in his orchard, his own bee population had more than doubled, from eight to seventeen healthy colonies, all producing delicious honey that he and his family ate regularly.[51] The editors of the *American Bee Journal* thought the claim preposterous and did not mind saying so. "Well! Well! We are astonished! When bees get fat, strong and healthy on Paris green! Who would have thought it? *Insect Life* [the journal in which the claim had been made] is published in Washington. We wonder if Prof. Wiley has not something to do with it."[e]

Opinion on the question was split so sharply that in 1891 the Association of Economic Entomologists established a three-man committee headed by F. M. Webster of Ohio and charged with determining once and for all if bees were killed by sprayed fruit blossoms. Webster's first report, issued a year later, related an experiment in which he had sprayed a blossoming plum tree with Paris green, enclosed the tree with mosquito netting, and then introduced a hive of bees inside the net.[52] Within fifteen hours, most of the bees had died, and chemical analysis showed their bodies

---

[e] *American Bee Journal*, 25, 644 (1889). The Professor was Harvey Washington Wiley, Chief of the Bureau of Chemistry in Washington and a man to be encountered later as the leading advocate of national pure food and drug legislation. In the early 1880s, Wiley's interest in food adulteration had prompted him to write several articles critical of the marketing of artificial honey made from glucose. Intended to chasten only an unscrupulous few, Wiley's remarks offended the entire honey industry, and for years apicultural writers looked for opportunities to take rhetorical revenge on "the Wiley lie." One such opportunity was found above.

had absorbed arsenic. For Webster, this was proof that arsenicals kill bees, but others were willing to believe that the bees in the experiment might have died from the trauma of unaccustomed confinement, or from the exertions of trying to escape the net. The diehard opposition was not converted until 1895, when Webster's final report described the detection of arsenic in the tissues of bees found dead in recently sprayed apple trees.[53] From this time forward, warnings not to spray fruit trees still in bloom became a standard part of entomologists' directions to farmers on the application of insecticides.[f]

The debate over bees and arsenicals illustrates the difficulty some agriculturalists had in recognizing that a substance that did them so much good could be guilty of doing others much harm. A similar blindness will become apparent in the later controversy over the hazards of arsenicals for people, but, in the early days of Paris green, many farmers were only too alert to the danger. Many years later, these days were recalled by the entomologist C. H. Fernald: "There was bitter opposition to the use of these insecticides for a long time, and the reports of cases of poisoning, which were said to have occurred at that time, were startling in the extreme. It was even claimed that potatoes would absorb the poison to such an extent that the tubers would carry poisonous doses, so that after each meal it would be necessary to take an antidote for the poison."[54]

As Fernald noted, these forebodings lasted a long time. As late as 1883, the *American Agriculturalist* reported that "We are but just recovering, in some of the rural districts, from the scare of using Paris Green,"[55] while a correspond-

[f] Insecticidal poisoning of bees has nevertheless continued to be a serious problem up to the present. In 1967, Clarence Benson, of the American Beekeeping Federation, estimated that ten percent of the country's five million bee colonies had been destroyed by pesticides during that year (cited by Graham, *Since Silent Spring*, p. 230). A columnist in *Newsweek* (Feb. 19, 1973, p. 92) has reported that the federal government allotted more than seven million dollars for 1973 to compensate beekeepers who lose bees to pesticides.

ent of the *Country Gentleman* related that a recent increase in green-skinned potatoes in his area was being attributed to Paris green poisoning.[56] The writer's own opinion was that the potatoes had been grown under too little soil and had been greened by sunlight, but his theory was being challenged by another local gentleman who was publicly calling for a boycott of Paris green treated potatoes and for legislation to prohibit use of the insecticide altogether. At this late date, a campaign to prohibit arsenicals was somewhat quixotic. Dangerous though the poison might be, it was the only thing that could satisfy the exigencies of insect control, and sooner or later most agriculturalists came to agree with a farmer named Allen, who "was quite afraid to use it [Paris green], until I found something must be done, or we would lose our whole crop, and until after trying a few rows, and watching the effect and making calculations of the chance of poisoning."[57]

The agricultural press encouraged this attitude by reiterating that "no other poison than arsenic in some form is effective, or applicable on the large scale."[58] Arsenic is surely dangerous, "as a sharpe axe is, and needs to be handled with care, but it should no more be tabooed on this account than the axe."[59] Caution and common sense, it was urged repeatedly, were the only protective measures necessary. Simply avoid inhaling or swallowing the arsenical dust when mixing or applying it, and no harm could follow. If uneasy over the popular rumors that potatoes treated with Paris green were poisonous, one need only consult the widely publicized findings of William McMurtrie or Robert Kedzie. McMurtrie had analyzed the pea plants grown in his arsenical soils (see p. 25) and found no traces of arsenic in any of them. Vegetables were therefore not poisoned by absorbing arsenic from the soil and, if McMurtrie's word failed to convince, there was the authority of Michigan physician Kedzie, who similarly had found no arsenic in potatoes taken from Paris green treated fields.[60]

These observations were most reassuring, but as the application of arsenicals was extended to crops other than potatoes, fears of poisoning developed anew. The treated produce was no longer underground, where it could be contaminated only by absorbing arsenic from the soil. Most fruits and vegetables were exposed directly to arsenical sprays, and had, in fact, to be thoroughly covered with the spray in order to be protected; but it would be a dubious kind of protection that saved the apple from the worm only by sacrificing it to arsenic. The question of the amount of arsenic remaining on harvested produce, the arsenical residue, thus assumed great significance for late nineteenth-century entomologists.[g]

Several men evaluated the possible hazard of spray residues experimentally, A. J. Cook being first to take to the field and foremost in his influence on other agriculturalists.[61] The Michigan entomologist carried out careful studies in 1880 in which he sprayed his fruit trees heavily with arsenicals and later subjected samples from the trees to chemical analysis. He found that nearly all of the applied arsenic was removed from produce within three weeks of application, by the action of wind and rain. One might, Dr. Cook concluded, eat as many sprayed apples a day as desired, so long as the apples had not been sprayed recently and if, just to be entirely on the safe side, they had been thoroughly washed and rinsed after picking. Cook's findings were widely publicized in the agricultural literature, as were those of subsequent students of the question, almost all of whom agreed that any health hazards from arsenical residues were remote. S. A. Forbes reported in 1887 that fruit that he had sprayed and then analyzed a

g It might be expected that after the introduction of lead arsenate, lead residues would become a subject of concern. Lead is quite poisonous in its own right, but little attention was given the possibility of poisoning by lead residues until the 1920s. Prior to that time, in America at least, it was generally assumed that any spray residue containing toxic amounts of lead must hold much more dangerous quantities of arsenic, and so arsenical residues attracted virtually all the attention.

week later carried as much as 0.9 mg. of arsenic per apple.[h] These were residues considerably greater than any found by Cook, but they were, after all, on produce that had been given only seven days "weathering," and that would still have to be consumed in extraordinary quantities ("74 apples would convey a poisonous dose") to cause injury. Somewhat lower, though still appreciable, levels of arsenic of the order of 0.01 grain per pound of fruit were soon to be found in other experiments allowing longer weathering periods,[62] while some investigations turned up no evidence of any residue on weathered produce.[63]

Disagreement over the amount of residual arsenic to be expected on harvested fruits and vegetables did not necessitate disagreement over the physiological effects of these residues. Rather, there was a general consensus among agricultural scientists, based on the residue studies cited above, that even the most heavily sprayed and unweathered produce was innocuous. This opinion overlooked the fact that those residue studies had been carried out under controlled conditions, that the produce selected for chemical analysis had been sprayed in most cases according to entomological recommendations for strength of spray and frequency of application, recommendations that were not necessarily followed outside agricultural experiment stations. American farmers have always been notoriously independent and when faced with a threat to their crops they might be expected to place greater faith in their own judgment than

[h] USDA *Annual Report*, 1887, p. 103. Different writers used different units of measurement in reporting weights of arsenic residue. Some, like Forbes, used the metric system, measuring the arsenic in milligrams (mg.), while others, the majority, employed the British system, expressing weights in grains. As the word implies, a "grain" is quite small, the equivalent of 1/7000 of a pound, or of approximately 65 mg. Nevertheless, a dosage of from one and one-half to five grains of arsenic suffices to kill most people, and even the seemingly miniscule level of 0.9 mg. per apple found by Forbes might produce chronic illness if consumed regularly over an extended period. The latter possibility, it will become clear, was not generally appreciated by entomologists.

in that of some experiment station theoretician presumably unacquainted with the practical hardships of husbandry. Entomological complaints of farmers spraying too heavily and too often were, in fact, not infrequent. One entomologist grumbled that although his state experiment station had never recommended more than three arsenical sprayings per season, it was a common practice among state farmers to spray as many as nine times![64] Another revealed that more than one farmer had personally confided his philosophy of spraying, "that if twice a month is good, four or five times as often would be a good deal better."[65]

Arsenical enthusiasm on the farm might have accounted for the frequent appearance in agricultural literature of reports of poisonings, some fatal, from the handling or eating of sprayed produce.[66] On the other hand, such reports might have been regarded as the imaginings of frightened, credulous rustics. The latter explanation was the one accepted by the agricultural press, which was inflexible in its denial that arsenical insecticides had ever caused any "authentic case of . . . injury." While acknowledging that "it is true a few individuals have claimed that they were made sick by eating sprayed fruit," agricultural writers affirmed that "in all such cases careful investigations have revealed that claims of this kind were absolutely without foundation."[67] These unsubstantiated poisonings were opposed by stories demonstrating arsenical insecticides to be safe in amounts much larger than might be encountered on produce. Entomologist Riley related the tale of "two negroes who stole some flour in which it [London purple] had been mixed in the ordinary proportion for use on cotton, and made biscuits thereof. Both were made sick but neither seriously, and Professor Barnard found that the steward on one of the Mississippi steamboats (the decks of which get quite purple from carrying it) has made regular use of the wastage so easily obtained on every hand for coloring his pastry and ice cream."[68]

Riley deserves recognition as perhaps the most cautious

of the entomologists in his attitude toward arsenical sprays. He concluded the above story with the comment that although "no ill results have followed is no reason for perpetuating this practice," and on another occasion admitted that his hopes that pyrethrum would be the "insecticide of the future" were motivated primarily by the desire to eliminate from use "the thousands of tons of Paris green, and other poisonous and dangerous arsenical compounds that have been sold all over the country."[69] Cook's study of arsenical residues finally persuaded Riley that Paris green could be used effectively and safely, though he was prompted several years later by Forbes's report of much larger residues to suggest that "fall poisoning possesses some danger on account of the cumulative effects of arsenic."[70]

Riley's vague suspicion of cumulative injury from spray residues is significant not so much because it was well-founded, as because it was virtually unique among agricultural authors. The true hazard of spray residues was indeed a chronic one, the danger of a slow deterioration of health caused by continued ingestion of small quantities of arsenic. As the following chapter will show, the question of chronic arsenicism was one of considerable interest to the medical profession of the day, as might have been appreciated by anyone who bothered to consult the profession's literature. Evidently few agriculturalists bothered. To the non-toxicologist, arsenic loomed as an acute hazard, as a poison that immediately caused serious illness or death. Preoccupied with the prevention of acute poisoning, entomologists and agricultural chemists naturally concluded that since spray residues never approached lethal doses, they must be harmless. "Even though all the poison sprayed upon the apples . . . should remain there undisturbed," a popular spraying handbook promised, "a person would be obliged to eat at one meal eight or ten barrels of the fruit in order to consume enough arsenic to cause any injury."[71] Similar blithe, but seemingly authoritative, assertions were issued regularly until well into the twentieth century. Agricultural writers delighted in calculations of the enormous numbers of sprayed apples or cabbages that would have to be eaten at a single sitting for a person to kill himself, and in implications that indigestion would bring on death long before arsenic began to exert its effect. Evaluation of the residue danger seemed a simple, even aphoristic matter: the editors of the *American Agriculturalist* summarized their profession's attitude with the observation that sprayed produce is clearly safe since "the proof of the pudding is in the eating, as we all know."[72]

"Eat arsenic? Yes, all you get,
Consenting, he did speak up;
'Tis better you should eat it, pet,
Than put it in my teacup!"

—Ambroise Bierce
The Devil's Dictionary

The Lingering Dram 2

BIERCE'S TALENT for disguising serious criticism as facetious poesy extended even to the realm of toxicology. Included among the noteworthy attitudes of his contemporaries in late nineteenth-century America was a strange ambivalence toward arsenic: a healthy fear of the most infamous of poisons was accompanied by a perverse willingness to accept it into the household in a variety of forms. Indeed, in some cases, arsenic was even eagerly welcomed and, in eating the poison, it will become apparent, Bierce's heroine was following the dictate of fashion as well as the advice of her husband. This confusion over the effects of arsenic, finally, extended into the ranks of practitioners of medicine, and its existence there goes far toward explaining the delay in the recognition of insecticide spray residues as a potential public health hazard.

Historically, medical interest in arsenic had been directed primarily toward its acute toxicity. Since the Renaissance, *la poudre de succession* had been resorted to so frequently by ambitious politicians and impatient relatives that physicians had perforce become expert in the recognition of acute arsenical poisoning. The appearance of nausea several hours after a meal, the progression to violent vomiting and

36

profuse diarrhea, followed by severe burning pains in the gastric region, intense thirst, difficulty in swallowing and speaking, and at last collapse and death after two to three days: this whole sequence of symptoms had become distressingly familiar. Doctors had further learned by the second half of the nineteenth century that arsenic did not have to be administered orally in order to poison, but could be absorbed through any of the mucous membranes, and even through external ulcers or the intact skin. To give the devil his due, the accumulation of this medical knowledge had depended nearly as much on imaginative criminals as on studious physicians. Toxicology texts of the period, in fact, often read like the *Police Gazette*, and famous murder cases were regularly cited to demonstrate the author's point. Among those recounted in discussions of arsenic was that of the French servant who at first failed to eliminate his lady employer with arsenical soup, but tried again and succeeded by adding arsenic to her enema syringe.[1] The introduction of arsenic into the vagina had been found by others to be equally effective and, as a combination of pleasure with business, proved considerably more popular. It was a method, furthermore, that need not have been restricted to the weaker sex, if the story of one Ladislas of Naples has any truth. According to the rumor repeated by nineteenth-century writers on poisons, the Neapolitan was brought to his end by a rival who concealed arsenic in the sexual organs of Ladislas' mistress. There was, finally, an even more deplorable route of administration of arsenic, the intoxication of the innocent babe through its mother's milk. In a notorious incident in France in the late 1880s, a husband was convicted of killing his year old child while attempting to poison his wife.[2]

Yet in the mid-1800s, the incidence of arsenical homicide was entering a decline, falling a victim to progress. The faster world of the nineteenth century fostered a desire for dispatch and efficiency that arsenic was unable to gratify. Poison looked old-fashioned beside the pistol. Chemical progress, too, discouraged would-be murderers from using arsenic. In the past, often the only evidence that a death had resulted from malicious poisoning had been the circumstances surrounding the case and the symptoms exhibited by the victim in his death throes. Both sets of data were subject to more than one interpretation, alternatives being food poisoning or accidental illness, and the murderer might reasonably hope that his crime would go unpunished, if not entirely unnoticed. During the early nineteenth century, however, analytical chemists developed techniques for detecting and estimating even traces of arsenic, and it became possible for suspicions of murder to be aroused by post-mortems as well as by circumstances. Arsenical murder was now too risky to maintain its previous popularity, and the poison came to be swallowed much more frequently by suicides than by unsuspecting victims.[a]

Arsenic was so frequently chosen for suicide because of its easy availability and reputation for thoroughness. For the majority of people, of course, the poison remained an object of horror, but the public, in its morbid fascination with acute arsenical poisoning, overlooked the more serious hazard, to which nearly all were exposed, of chronic arsenicism. The relatively large lethal doses administered by the poisoner or swallowed by accident could, after all, be easily avoided by all but the careless and unfortunate. But arsenic was also a common ingredient in a wide variety of manufactured items, and many people were regularly exposing themselves to quantities of arsenic that, though

[a] Sensational cases of arsenic poisonings still occurred. As late as the 1880s, a Dutch woman was sentenced to life imprisonment for having given arsenic to 102 people, of whom 27 died and 47 were made seriously ill. Arsenic poisoning was especially popular during the cholera epidemics which prevailed during the nineteenth century. The similarity between the symptoms of cholera and those of acute arsenical intoxication suggested that in the panic of the epidemic, poisoning victims would be mistaken for cholera patients. The fact that we know of this supposition demonstrates that at least some of the murderous opportunists were caught.

too small to produce immediate illness, might result in long-term cumulative damage.

The sources of exposure were so many that a listing of arsenically contaminated products alone might validate the historical cliché that *caveat emptor* was never better advice than during the late nineteenth century. Arsenical pigments, particularly Scheele's green (cupric arsenite), were too cheap and brilliantly colored to escape the manufacturer's notice, and they were employed freely in all European countries and the United States. An incomplete list of arsenic-tinted items compiled by the Medical Society of London in the early 1880s, for example, enumerated "paper, fancy and surface coloured, in sheets for covering cardboard boxes; for labels of all kinds; for advertisement cards, playing cards, wrappers for sweetmeats, cosaques, etc.; for the ornamentation of children's toys; for covering children's and other books; for lamp shades, paperhangings for walls and other purposes; artificial leaves and flowers; wax ornaments for Christmas trees and other purposes; printed or woven fabrics intended for use as garments; printed or woven fabrics intended for use as curtains or coverings for furniture; children's toys, particularly inflated india-rubber balls with dry colour inside, painted india-rubber dolls, stands and rockers of rocking-horses and the like, glass balls (hollow); distemper colour for decorative purposes; oil paint for the same; lithographer's colour printing; decorated tin plates, including painted labels used by butchers and others to advertise the price of provisions; japanned goods generally; Venetian and other blinds; American or leather cloth; printed table baizes; carpets, floorcloth, linoleum, book cloth and fancy bindings. To this list may be added coloured soaps, sweetmeats and false malachite. Arsenic is also used in the preparation of skins for stuffing and of some preservatives used by anatomists."[3]

Other lists, though less lengthy, often turned up additional arsenical products. The renowned British toxicologist Robert Christison observed that not only was Scheele's green used to make sweetmeats more appetizing, but that it was also added to preserves and to apple tarts, and that several children had been made ill by the latter.[4] Others[5] pointed out that the green cakes in water color sets generally contained arsenic, as did dental fillings, and that people had been injured by arsenical stockings, veils, cosmetics, concert tickets, fly papers, stuffed animals, even money. There were rumors that clerks employed at the U.S. Treasury to count bank notes often suffered from skin eruptions caused by arsenic in the notepaper,[6] while at the other extreme it was discovered that arsenic had infiltrated the kindergarten, gaining entrance in the brightly colored papers so favored by children for artwork.[7] From the Temple of Mammon to the Halls of Learning, arsenic lurked everywhere, though its presence seems to have been most menacing in candles and wallpapers. Arsenicals were used not simply to color wax candles, but were given employment in the 1830s in the production of new "composite candles." These were mixtures of tallow with arsenic compounds that gave the tallow the texture and appearance of more expensive wax. Medical attention was quickly called to the arsenic content of the candles, and one writer even renamed them "corpse candles."[8] Nevertheless, arsenical candles remained on the market and, as late as the closing year of the 1800s, several children were poisoned by green candles on a Christmas tree.[9]

Candles inflicted only minor casualties, though, compared to the victims claimed by wallpapers, products capable of poisoning the air of the rooms they lined by releasing both the original pigment, in the form of a fine dust, and the gas, ethyl arsine $(As(C_2H_5)_2H)$, produced by the action of certain fungi on the pigment. Several European and American physicians had spoken against the sale of arsenical wallpapers as early as the decade of the 1860s and by the 1880s their complaints had swelled to a chorus. The conclusion that wallpapers could be hazardous was indicated by the growing incidence during the second half of the century of patients with symptoms of mild arsenic poisoning—lassitude, headache, irritated mucous membranes, skin irritation, and arsenic in the urine—whose conditions cleared up when they were transferred to new surroundings. When these same patients suffered relapses on returning to their homes, it seemed clear their ailments had been caused by some arsenical product in the original environment. In spite of the seeming ubiquity of arsenic in commercial goods, on occasion the only arsenical product to be found in the patient's house was his wallpaper, and enough papers were incriminated in this way to excite a vigorous agitation against arsenical wallpapers, as well as other manufactures, during the closing decades of the century. Several state boards of health investigated the problem, and the Michigan board even prepared a book composed of pages of wallpapers from leading dealers in the state. Entitled *Shadows From the Walls of Death, or Arsenical Wall-papers*, the volume was distributed to 100 public libraries, at one of which, it was reported, a lady was poisoned by simply thumbing through a copy.[10]

The real leadership of the campaign against arsenical manufactures was assumed by the physicians of Massachusetts, most prominent among these Yankee meddlers being a foursome of Bostonians: Frank Winthrop Draper, medical examiner (coroner) of Boston and lecturer at Harvard Medical School; James Jackson Putnam, professor of neurology at Harvard; Frederick Cheever Shattuck, professor of clinical medicine; and William Barker Hills, the medical school's chemistry professor. With a diligence befitting Harvard men, these four delivered lectures and published papers on the dangers of arsenic in the household, with the express purpose of informing the citizenry, through their physicians, of products to be avoided. It was optimistically hoped that consumers could be educated to demand arsenic-free merchandise, and that the laws of the marketplace would then force manufacturers to abandon arsenical pigments, though it was appreciated in more realistic moments that even an enlightened public could not be relied upon to protect itself and that state intervention, in the form of prohibitive legislation, would be desirable. It would also, it was realized, be difficult to obtain. Laws restricting the quantities of arsenic in marketed merchandise had been in effect in most European countries for years, yet Americans had shown no inclination to follow the Old World example. Draper suggested with cynical confidence that such an inclination would not soon appear, either, because in America, where "liberty runs mad . . . the rights of individuals and of industrial pursuits are deemed too sacred to allow of excessive restriction."[11] Draper himself nevertheless initiated the move to impose restrictions on arsenic in an article exposing "the evil effects of the use of arsenic in certain green colors,"[12] included in the annual report of the Massachusetts State Board of Health, presented to the state legislature in 1872. As anticipated, American liberty was not to be easily infringed. Draper's recommendations for arsenic-control legislation were ignored, but undaunted, and with medical reinforcements, he maintained a siege of the state-house until its occupants finally capitulated in 1900 with "An Act relative to the manufacture and sale of textile fabrics and papers containing arsenic."[b]

In the interim, several similar bills had been proposed and voted down, being defeated, according to their medical sponsors, by the "determined and well-organized opposition on the part of the manufacturers and dealers."[13] Industrial opposition to these bills was undoubtedly most strenuous,

[b] The Act, reprinted in *Boston Medical and Surgical Journal*, *143*, 413 (1900), established limits of 0.01 grain of arsenic per square yard for fabrics used in articles of dress, and of ten times that concentration for other articles. Fines of fifty to two hundred dollars were provided for offenders. This was the first state law aimed at controlling arsenic in manufactures, but other states eventually adopted similar legislation.

yet the objecting doctors should have admitted that a part of the blame for the delay in adopting arsenic legislation had to rest with themselves. Convinced as they were that domestic arsenic was dangerous, physicians still experienced great difficulty moving politicians to the same conviction. The hazard of arsenic, after all, was not one easily appreciated by non-toxicologists. Wallpapers and other contaminated items usually contained too little of the poison to cause immediate illness, and the suggested possibility that less dramatic, but nonetheless serious, symptoms might develop gradually after months of exposure must have seemed a fanciful hypothesis to many laymen. Industrialists certainly doubted that their products were poisoning customers: a paper salesman named Bumstead informed an 1887 meeting of the Massachusetts Medical Society that "the paper-dealers do not place credence in very many of the alarming reports which are . . . circulated in relation to the occurrence of dangerous interference with health from the action of the colors used in wallpapers."[14] The paper-dealers' analysis of the question, of course, was weighted by economic considerations, but even the most objective legislator might have shared their skepticism: the hypothesis of an epidemic of chronic arsenicism was supported by circumstantial evidence—the appearance of certain symptoms in patients, the discovery of arsenic in the patients' wallpapers, and the detection of more arsenic in their urine—and the doctors themselves admitted that this concatenation was "not absolute proof that the arsenic came from the wall-paper" or that it had caused the symptoms.[15]

More damaging to the anti-arsenical cause was the problem of dissension within the medical ranks. The proposition that arsenic-containing wallpapers posed a public health threat did not come close to commanding a professional consensus: the director of the Health Department of New York City,.for instance, proclaimed that he had "carefully considered the whole subject of arsenical wallpapers

... and concluded that there was nothing in it."[16] In view of such contrasting medical opinion, the sluggishness of the Massachusetts legislature in moving against arsenical wallpapers and fabrics is understandable, and one is almost surprised they moved at all.

This matter of medical disagreement over the seriousness of the arsenic danger demands further analysis since it affected not simply the specific hazard of Massachusetts wallpaper, but more importantly the problem of general arsenical contamination of the entire country, including the contamination of produce with insecticide residues. The sources of medical disagreement were several, and not entirely of a purely scientific nature. Most physicians did agree by the late nineteenth century that chronic arsenical intoxication could occur. The general concept of slow poisoning was, in fact, a very old one in medicine. It had been accepted as a matter of course in antiquity that some poisons could be administered so as to cause gradual debilitation rather than sudden death, and more recently the idea had attained renewed popularity from the Renaissance reverence for the knowledge of antiquity. Camillo's recourse to

> ... *no rash potion,*
> *But ... a ling'ring dram, that should not work*
> *Maliciously like poison.*
> (The Winter's Tale, Act I, Scene II)

reflects this belief, and an infamous seventeenth-century lady poisoner named Tofana was supposedly skilled at concocting arsenical drugs that might bring a victim "to a miserable end through months or years, according to the enemy's desire."[17] "L'Affaire des Poisons"—that blending of witchcraft and drug-lore with murderous conspiracy which so unsettled the court of Louis XIV—further illustrates the fear of slow poisoning. In the 1682 edict restricting the use of poisonous substances, frightened out of Louis by the affair, special prohibition was made of those poisons, left

44

unidentified, "which work more slowly, causing gradual deterioration or lingering illness."[c]

Neither Shakespeare nor Tofana nor even the glorious Louis were recognized as authorities by late nineteenth-century toxicologists, but slow intoxication was becoming generally appreciated. Experience had indicated lead and mercury to be the worst offenders, but it was also known that continued exposure to far from lethal amounts of arsenic might lead to irritations of skin and mucous membranes, and to nervous disorders. The bones of contention were the questions of how frequently such poisoning occurred and how much arsenic ingested over how long a time was necessary to produce it. The general opinion was that chronic arsenicism was rare, affecting only the specially susceptible exposed to relatively high levels of the poison (two to three grains of arsenic was the acknowledged lethal dose, and significant fractions of a grain might be considered "relatively high"). There were a few, however, who suspected arsenical illness might be much more common. Among these, J. J. Putnam was the most voluble. The Harvard neurology professor had pioneered in his subject in the United States, had founded a neurological clinic at Massachusetts General Hospital, and had devoted a great deal of attention to the subjects of lead and arsenic poisoning. These experiences had convinced him that instances of arsenic poisoning occur "more commonly than has been supposed, but pass often unobserved, partly because this diagnosis is not in the physician's mind, and partly because the symptoms are often obscure and attributable to other causes."[18] Putnam was aware that arsenic could produce neurological damage and that the sources of arsenic in the environment were increasing during his time, so at least in his mind the diagnosis of chronic arsenic poisoning was

---

[c] Quoted in Frances Mossiker's *The Affair of the Poisons*, New York, 1969, p. 276, a scholarly investigation into "one of history's great unsolved mysteries," and a marvelously entertaining account of the dalliances and dangers of life with the sun king.

45

established as a distinct possibility. It became even more distinct in 1894 when his colleague in chemistry, W. B. Hills, reported that of 180 patients suspected by Putnam to be suffering from chronic arsenicism, 75 percent were excreting arsenic in their urine.[19] Putnam was probably not surprised, since he had been informed by another chemist three years before that, of 150 patients selected at random, more than 30 percent were excreting appreciable quantities of arsenic.[20]

With so many people obviously absorbing arsenic from their surroundings, it seemed only logical to Putnam that some should be adversely affected, and he urged his fellow physicians to respect chronic arsenicism as a potentially common diagnosis. His further intent was to educate them to recognize cases of chronic poisoning when they saw them. As early as 1888, Putnam had reported on patients whose chronic arsenic poisoning had been so serious as to progress to paralysis, and even death,[21] but it was the less traumatic cases that were his special concern, the cases in which "the symptoms are . . . obscure and attributable to other causes." Multiple neuritis, an inflammation of nervous tissue accompanied by pain and losses of sensation and reflex activity, he had found to be "of the most common occurrence,"[22] though frequently overlooked by patients and doctors alike. Even less likely to attract attention were milder, generalized symptoms such as anemia, digestive disturbances, or local neuralgia, yet patients suffering from all of these had been found to be excreting arsenic. Putnam readily admitted that his evidence was presumptive, that it was sufficient to incriminate arsenic, but not to convict it of mass poisoning. His position (and that of his fellow arsenic critics) was that conclusive proof of widespread arsenicism might be unobtainable, and that in such a situation prudence dictated that arsenic be assumed guilty and removed as completely as possible from the human environment.

The medical profession nevertheless remained generally

46

skeptical of the arsenic hazard, and for reasons additional to the difficulties of tracing minor ailments to small amounts of a common poison. Not the least of these was the fact that arsenic was a time-honored medicine still in wide use. Historically, substances have rarely been refused admission to the materia medica simply because of their potential toxicity. Particularly after the resolution of a protracted seventeenth-century altercation between the proponents of traditional herbal remedies and those of recently discovered mineral medicines, physicians came to prescribe dangerous metallic compounds with a generosity difficult to credit today. Theoretical support for these remedies came from the swarm of speculative medical systems developed during the eighteenth century, systems that, each in its own way, reaffirmed the ancient belief that depletive procedures such as bleeding, purging, and vomiting were the surest exterminators of disease.[23] Consequently, such drugs as the purgative calomel (mercurous chloride) and the vomitive tartar emetic (antimony potassium tartrate) were administered so frequently and liberally as to earn the epithet "age of heroic therapy" for the century preceding 1850.

There was also apparent by 1850, however, a revolt against heroic drugging. The "Paris school" of clinicians began early in the century urging the rejection of the hypothetical systems that sanctioned evacuative therapy, in favor of a new medicine to be derived from the correlation of data collected in careful clinical and post-mortem studies of large numbers of subjects. The process of reform would admittedly require time, including an interim during which physicians would have to be preoccupied with accumulating the information required for a correct understanding of the nature of disease. Without such an understanding, therapy could not be rationally directed, and there thus came to be associated with the Paris school a philosophy of "therapeutic skepticism," the attitude that, at least for the present, nature is the best doctor; that the doctor should abandon the aggressive approach of traditional medicine

and restrict his activity to keeping the patient rested, nourished, and as free as possible from pain and care.

Converts to therapeutic skepticism were especially critical of the continued administration of harsh mineral remedies. A New York physician (and ever-competitive America
was naturally the most medically heroic of nations) spoke
for the skeptics when he proposed that the great virtue of
calomel was that: "If there is nothing the matter with the
person who takes it, there very soon will be; and although
before its administration, it might be impossible to know
or say what was the matter—if anything—it will be very
easy to do both, after it has been given. . . . If a medical
man cannot find enough of disease to employ him, let him
give calomel to that which he does find, and he will most
assuredly find more."[24] The public agreed. A popular ditty
of the 1830s and 1840s recited the dire effects of the drug,
and closed with a plea to heavy dosing doctors:

> And when I do resign my breath
> Pray let me die a natural death,
> And bid you all a long farewell,
> Without one dose of calomel.[25]

Calomel's dominant position in the old therapy rendered
it the chief point of assault by the skeptics, but all drugs
came under attack to a degree. Oliver Wendell Holmes'
suggestion that the sinking of the materia medica to the
bottom of the sea "would be all the better for mankind,—
and all the worse for the fishes," was a deduction of medical
common sense. As Holmes innocently observed in an 1861
(pre-Paris green) address to Harvard medical students, "the
farmer would be laughed at who undertook to manure his
fields or his trees with a salt of lead or of arsenic"; the same
response should be given the physician who strives to improve his patient's health with toxic "drugs."[26] In 1861,
Holmes was beating a dying horse, for heroic therapy was
well along in its decline, but at least one of the old remedies
still had a few good days left.

Compounds of arsenic have been given varying medicinal uses since antiquity, but it was not until late in the eighteenth century that the drug entered into its career as a near panacea. In 1781, a London apothecary named Wilson was granted a patent for his Tasteless Ague and Fever Drops, a potion that seems to have sold well throughout the realm but that was employed with most notable success by Dr. Thomas Fowler, physician to the Stafford infirmary.[27] Learning from the infirmary's chemist, Mr. Hughes, that the drops contained arsenic, Fowler concocted his own remedy of potassium arsenite solution. His 1786 *Medical Reports of the Effects of Arsenic in the Cure of Agues, Remitting Fevers, and Periodic Headaches* established Fowler's solution" as a therapeutic mainstay, and within a century it was being regularly prescribed for many ailments in addition to the original agues, fevers and headaches. The reason for arsenic's meteoric rise in medical favor was the simple, but very good, one that it seemed to work. Taken in the proper amounts, arsenic often exerts an apparent tonic effect, producing a feeling of increased vigor, a gain in weight, and a smoother, rosier complexion. These improvements are generally only temporary, being early manifestations of the poison's action on the circulatory system, but appearances can be deceiving: in the case of Fowler's solution, apparent success with one ailment led to its recommendation for another, until it was possible for a physician of the 1890s to observe that arsenic had come to occupy a pedestal in therapeutics.[28] The respect in which Fowler's solution was held can be best appreciated by an enumeration of the conditions for which it was regularly prescribed. These included anemia, headache, dyspepsia, eczema, psoriasis, all other chronic skin diseases, neuralgia, chorea, epilepsy, whooping cough, asthma, bronchitis, emphysema, pulmonary tuberculosis, malaria, and cancer. In addition, Fowler's solution was suggested, at one time or another, for cholera, yellow fever, syphilis, diabetes, angina pectoris, tic douloureux, gout, arthritis, rheumatism, constipation, morning sickness, melancholia, impotence, fits of sneezing, warts, boils, and, for the careless traveler in the Orient, cobra bites. Arsenic was further widely regarded as a beneficial addition to calomel purges, nor does this recitation exhaust the possibilities for use of the potent drug. There were yet more ailments presumed susceptible to it, and it might be administered in more than one way for each. The oral route of administration was the one most frequently taken, but hypodermic injection was an occasionally used alternative, while inhalation of arsenical fumes was especially indicated for respiratory conditions. A Philadelphia physician of the 1890s advised his respiratory patients to smoke daily two or three cigarettes previously dipped in Fowler's solution, then dried.[29]

Propriety excluded the fairer sex from this last form of therapy, but they had their special compensations. The fame of arsenic as a specific for skin disease had suggested to several entrepreneurs that it must be invaluable for skin hygiene as well, and an array of arsenical cosmetics were marketed with the promise that "your complexion can be made to rival the lily and the rose."[30] Claims like this impressed even physicians as carrying a good thing too far, and commercial puffery was at least partially offset by medical denunciation. "What a prospect is this," one doctor fumed, "for the man whose wife is thus absurdly immolating herself on the altar of vanity! What danger it intimates for the infant whose nurse shares in this ambition to be beautified!"[d]

It was, naturally, non-prescription arsenic that incurred the greatest medical wrath. When the object of treatment was restoration of health rather than enhancement of beauty, physicians were considerably less critical of the administration of the drug, with the result that it is quite

[d] *The Medical and Surgical Reporter*, *39*, 194 (1878). Ambrose Bierce agreed, if with less outrage, by defining arsenic as "a kind of cosmetic greatly affected by the ladies, whom it greatly affects in turn" (*The Devil's Dictionary*).

plausible that one of the ailments for which Fowler's solution was most frequently administered was chronic arsenical poisoning. Arsenicism was not a common diagnosis, but probably a common occurrence, and the greatest single source of such poisoning was perhaps the medicine bottle. With Fowler's solution being prescribed so freely, during a period when chronic arsenicism was but little appreciated, it would indeed be surprising had there not been a high incidence of iatrogenic arsenic poisoning. John Winslow has recently presented an appealing argument that much of Victorian England's notorious gastric and nervous distress, previously designated dyspepsia, was in reality "Fowler's disease."[31] His diagnosis was of Charles Darwin in particular, and of Darwin's countrymen in general, but it is just as applicable to Americans of that day. Dyspepsia was no less common here (more than one writer of the period referred to it as "Americanitis"), and many American physicians yielded place to no one in their loyalty to Fowler's remedy.

Fortunately, not all the doctors were so distracted by arsenical zeal that they overlooked the harm often done by the drug. There had been occasional criticisms of the use of arsenical medication since Fowler's time, but these became increasingly numerous during the last quarter of the nineteenth century.[32] Among the ailments attributed by critics to overdosing with arsenic were gastric distress, coughing, headache, peripheral neuritis, paralysis, and skin disorders such as keratosis and melanosis. Above all these dangers, though, hovered the specter of cancer, the most feared of modern diseases and one that during the late 1800s was beginning to be associated with exposure to arsenic.

That particular chemical substances might gradually induce cancer had been first suggested a century before, by the prominent British surgeon, Percival Pott. In 1775, Pott announced that the exceptional frequency of cancer of the scrotum among chimney sweeps was a consequence of prolonged, nearly constant contact with chimney soot.[33] A very similar charge was directed against arsenic almost half a century later by the London physician, John Ayrton Paris, in the first edition of his much published *Pharmacologia*. Recalling his earliest years of medical practice, in the mining and smelting district of Cornwall, Paris claimed to have observed "the pernicious influence of arsenical fumes [from the copper smelteries] upon organized beings," and that, as a result of exposure to these fumes, "horses and cows commonly lose their hoofs, and the latter are often to be seen in the neighboring pastures crawling on their knees and not infrequently suffering from cancerous infection in their rumps. . . . It deserves notice that the smelters are occasionally affected with a cancerous disease in the scrotum similar to that which infest chimney-sweepers."[34] The *Pharmacologia* was but one of Paris' several claims to scientific fame; the others ranged from the lending of his name to a certain green arsenical pigment to the presidency of the College of Physicians of London. A man of such stature was not to be challenged lightly and, understandably, his incrimination of arsenic in scrotal cancer went unquestioned for quite some time. Not until 1892, in fact, was the incidence of cancer in Cornwall carefully investigated. The investigator, H. T. Butlin, interviewed all the Cornish physicians he could locate, and discovered that not one had ever treated a case of scrotal cancer among the local smelter workers.[35]

Paris had apparently exaggerated the experiences of his youth, but his discrediting came too late to free arsenic of the indictment of carcinogen. Several years prior to Butlin's interviews, yet another English doctor, Jonathan Hutchinson, had substantiated a suspicion that had been growing for more than a decade that the arsenic administered so liberally to relieve skin conditions like psoriasis might, instead, aggravate them to the point of producing skin cancer. Hutchinson cited five of his own cases in support of this contention, cases that were subsequently increased by the reports of so many other practitioners that, by the 1920s, "the views of Hutchinson were accepted by all dermatologists."[36]

Thirty or forty years earlier, Hutchinson's distrust of arsenical medication had not been so universally shared. For every physician who feared that repeated administration of Fowler's solution might lead to chronic intoxication, if not cancer, there were colleagues who were confident that only doses considerably larger than medicinal ones could cause any harm. The frustrations of the campaign against wallpapers were repeated: the case against arsenical medicines, though persuasive, was not so conclusive as to preclude reasonable doubt. Fowler's solution had the accumulated endorsements of several generations of physicians; its salutary effects could be clearly seen in patients suffering from a variety of ailments, while its alleged injuries were so non-specific as to be attributable to any of several factors. The desire to believe in the beneficence of arsenic in moderation, furthermore, was encouraged not only by the experiences of medical practice but also by the lore of medical mythology, in particular by the numerous reports of a race of exceptionally healthy men and women, the arsenic-eaters of Styria.

The ore-rich Austrian province of Styria had once been a duchy in the Holy Roman Empire. The early-nineteenth-century decline in their political status, however, did not mean any loss of fame for the Styrians, for, contemporaneous with the demotion from duchy to province, Styria became the focus of considerable excitement among medical scientists throughout Europe. The cause of the excitement was the so-called toxicophagi, the poison-eating peasants who spiced their otherwise humble fare with lethal portions of arsenic—at least, the quantities of arsenic they ate were lethal for lesser men. The Styrians thrived on their bizarre dietary supplement, and were convinced that it gave them robust health and longevity. This faith in arsenic as an essential of hygiene was doubtless suggested by the same

53

effects (increased vigor, weight gain, improved complexion) that had recommended it as a medicine, though the Styrian peasantry surely discovered these effects for themselves, without benefit of medical guidance.[e] It was not, in fact, until the publication of the sensational reports of Dr. Johann von Tschudi in the early 1850s that this practice of toxicophagy became generally known to the medical profession. Though originally published in the author's native language, his reports contained such startling revelations that they were quickly translated from German and reprinted in the leading medical journals of Europe and the United States.[37] On the heels of von Tschudi followed popularizers such as the British chemist James Johnston, who informed readers of his *Chemistry of Common Life* that Styrians consumed arsenic "first, to give plumpness to the figure, cleanness and softness to the skin, and beauty and freshness to the complexion. Secondly, to improve the breathing and give longness of wind, so that steep and continuous heights may be climbed without difficulty and exhaustion of breath. Both these results are described as following almost invariably from the prolonged use of arsenic by men or by animals."[38] Embellishments added by other writers included claims that this wondrous drug stimulated both mental activity and the sexual appetite, the latter effect accounting for the rumored exceptional rate of illegitimate births in Styria.[39]

More remarkable than all these virtues of arsenic was its apparent harmlessness when taken according to the Styrian custom of gradually increasing dosage. It was the supposed ability of the body to habituate itself to a normally deadly poison that created a medical sensation, but although the

[e] The undocumented discovery of the benefits of arsenic for people was made, most likely, either by a farmer (cattle and swine throughout Europe and America had for years been given small portions of arsenic with their rations to accelerate the fattening process) or by a smelter worker (arsenic was a common byproduct of the many Styrian smelteries).

Styrians often took huge amounts of arsenic, in effect their doses of poison were much smaller. It was eventually pointed out by several pharmacologists, and most notably by E. W. Schwartze at the U.S. Department of Agriculture (1922),[40] that although Styrians might gulp down several grains of arsenic at a time, they swallowed the poison in dry lump form. In such form, only a portion of the arsenic would dissolve and be absorbed from the gastro-intestinal tract, while the remainder, the bulk of that ingested, would soon be excreted with the feces. If the quantities ordinarily eaten were to be taken in solution, death—not habituation—would be the result.

Schwartze's skepticism was nothing new. Incredulity had also been the reaction of most of von Tschudi's readers over half a century before. At that time, habituation to poisons was a recognized phenomenon, but it was believed to be confined to drugs of organic origin, such as opium and alcohol. The toxicological authority of the period, Robert Christison, felt quite certain that "the inorganic poisons are most of them little impaired in activity by the force of habit. . . . There is no satisfactory evidence that a person, by taking gradually-increasing doses of arsenic, can acquire the power of enduring a considerably larger dose than when he began."[41] Christison went on to reject the rumors of arsenic and corrosive sublimate (mercuric chloride) eaters that had come to his attention, and when, only a few years later, such rumors gained currency again, many medical men naturally dismissed them. A typical response was that of the British physician who considered that "the story of the Styrian arsenic-eaters is not only unsupported by adequate testimony, but is inconsistent, improbable, and utterly incredible."[42]

If most physicians echoed this refrain in the beginning, the tune of many soon changed, as "adequate testimony" began to be provided. Several curious physicians, British as well as Austrian, conducted on-the-scene investigations of arsenic-eaters, and concluded that von Tschudi's reports had not been exaggeration after all. Most convincing of the substantiations of toxicophagy was that of Craig Maclagan of Edinburgh. Dr. Maclagan had experienced the usual trouble locating arsenic-eaters for investigation; the habit, he had quickly discovered, was one of those pleasures which people privately enjoy but publicly condemn, and few Styrians could be persuaded to admit their arsenical intemperance. The persevering Dr. Maclagan did finally find two cooperative toxicophagi, observed them swallowing what should have been lethal doses of arsenic, verified by analysis that the swallowed substance was truly arsenic, and, by further analysis, detected arsenic in his volunteers' urine. Throughout these proceedings, Maclagan's subjects displayed not the least sign of discomfort, except for a mild belch emitted by one shortly after his arsenical meal, a belch dutifully recorded by the thorough Maclagan before he dismissed it as insignificant. The doctor concluded that human habituation to arsenic had been confirmed and that toxicological doctrine would have to be revised.

Maclagan's and other scientific eyewitness accounts of arsenic habituation dispelled much of the medical profession's skepticism about arsenic-eating, though there remained physicians who refused to swallow such reports without at least one grain of salt. Von Tschudi himself had admitted that not all Styrians who adopted the arsenic habit continued it unharmed: among many cases that had come to his knowledge, he reported, was that of a dairy maid who, in pursuit of beauty, took too much arsenic and "fell a victim to her coquetry [and] died poisoned."[43] Later investigators were to add still more cases of poisoned Styrians to the medical record, cases of chronic as well as acute arsenicism,[44] and instances of non-Styrian arsenic-eaters poisoning themselves came to light on occasion. In 1864, for instance, a Nova Scotia man who had read about the Styrians' strange custom four years earlier in a popular magazine died from the effects of adopting the practice as his own.[45] Although able to survive daily two to three grain doses, he had suffered several years from dyspepsia, sore throat, and darkening skin, all signs of chronic arsenical poisoning. He had nevertheless persisted in his habit for the understandable reason that it seemed to stimulate his libido, and his neighbors readily volunteered that "for a long time he had been notorious for his amorous propensities." He unhappily learned too late that Eros cannot be worshipped to the neglect of Hygeia.

This last arsenic-eater's history also suggests the ease with which nineteenth-century laymen might be led astray by non-medical discussions of the Styrians. Popular articles tended to accentuate the positive aspects of arsenic-eating, and those who read magazines rather than medical journals could easily conclude that small amounts of arsenic were only beneficial. It will be seen that agriculturalists sometimes drew such conclusions, and called in the Styrians as reinforcements for their defense of the safety of insecticide residues.

The foregoing digression on arsenic and the medical profession in the late nineteenth century illuminates the period's indecision over whether or not environmental arsenic posed a public health threat. The many reported cases of illness in inhabitants of arsenical surroundings surely provided good reason to suspect that low levels of arsenic might be hazardous. On the other hand, the vagueness of the symptoms and the inexperience of physicians in observing them dictated that many cases of chronic arsenicism might pass undetected, while in the cases that were detected, the circumstantial nature of the evidence against arsenic made conviction of the poison uncertain. Indeed, a jury of American physicians would have been disinclined, by education and experience, to even suspect arsenic as the culprit in cases of chronic illness. Reputable foreign medical scientists had observed that Styrians, and therefore presumably other people, could actually benefit from regular exposure to arsenic, while native doctors had employed Fowler's solution freely, and with apparent success, for years. Suspicion might more naturally be directed at the critics of arsenic than at the drug itself, and it might seem especially telling that the most vocal of these critics were residents of Boston.

Often praised as the Athens of the New World, frequently chided as the hub of Yankee eccentricity, Boston was the split personality among nineteenth-century American cities. This characterization is especially applicable to Boston medicine, for the city where inoculation against smallpox was introduced, and where the wonder of ether anesthesia was first demonstrated, was also peculiarly prone to medical faddism. The phrenologists, homeopathists, hydropathists, mesmerists, and others among the hordes of unorthodox medical men who swarmed about the America of the mid-1800s nowhere found a warmer welcome than in Boston. Regular Bostonian practitioners were, of course, painfully aware of the seemingly hereditary injudiciousness of their fellow citizens, and one remarked in 1849, only half in jest, that "if some distinguished quack should announce in Boston that covering the nose with green paint would prove a specific for the whole family of human diseases, painted noses would be in a triumphant majority within a month."[46]

Forty years later, it was green wallpaper, not noses, that was the object of medical interest in Boston, but the city was still regarded as the "city of notions." Consequently, when Boston physicians began to exhibit such anxiety over a matter that had aroused only occasional interest among the doctors of other areas, it was easy for the anti-arsenic campaign to take on the appearance of the latest "Boston fad." Harvard's Professor Shattuck soon found it necessary to answer "the objection that arsenical poisoning of domes-

tic origin occurs only in Boston,"[47] though even his elo-
quent deflating of the facetiousness failed to halt it. In
mock alarm, the New York City editors of the *Medical
Record* alerted their subscribers to "the condition of our
esteemed and interesting colleagues of Massachusetts."[48]
"Having passed safely," the editors continued, "through
disquieting exacerbations of homeopathy, mind-cure, Ibsen-
ism, and psychical research, the profession has been brought
up sharply with an attack of arseniophobia . . . the evidence
seems to accumulate in favor of the view that in Boston
mural arsenic poisoning is an infectious disease. . . . In fact
there seems to be an appalling possibility that Massachusetts is being systematically poisoned by an inoculable,
malignantly infective, and extremely prevalent form of
arsenical poisoning. We rejoice to learn that the Legislature
and State Board of Health are at work upon the matter."

A crusade so widely mistaken for a craze could not expect
to win many converts, and the Boston-led campaign against
arsenic predictably fell short of the goals of its directors.
The Massachusetts legislature finally consented to regulate
the arsenical content of wallpapers and fabrics, but both
the legislators and the physicians of other states were more
obstinate. Less than a year before enactment of the Massachusetts wallpaper law, a contributor to a prominent medical journal had commented that "the conclusion to be
drawn [from the frequent occurrence of arsenic in wall-
paper] is obviously that it is inexpedient to eat wallpaper
unless one happens to be suffering from malaria or is in
need of a strong tonic."[49]

The potential dangers of an arsenical environment were
clearly not a matter of universal alarm, and the specific
hazard of insecticide residues produced the least stir of all.
The scientific literature of the period, in fact, contains only
two significant warnings by medical writers to avoid
sprayed produce. William Hills, chemistry professor at
Harvard Medical School, regularly examined urine specimens for the presence of arsenic and shared his findings

with his comrades in the campaign against arsenical papers. In 1894, Hills found a patient who was excreting large amounts of arsenic but had had no exposure to any of the usual domestic sources of the poison. Since the man's urine also contained appreciable quantities of copper, Hills suspected that his illness might be the result of Paris green residues and so cautioned that "we may have, in the free use of Paris green in the field and garden, one explanation of the frequent occurrence of arsenic in the system."[50]

Another physician-chemistry professor, Robert Clark Kedzie, was much more deeply involved in the residue question. From his faculty post at Michigan Agricultural College, Kedzie had promoted "farmers' institutes" throughout the state, institutes devoted to increasing agricultural prosperity by the aid of applied chemistry. The particular application of chemical insecticides had been sanctioned by Kedzie through his demonstration (cited in Chapter I) that potatoes do not absorb arsenic from the soil. It was Kedzie, also, who had carried out the analyses for arsenic used by A. C. Cook in his 1880 defense of sprayed fruit. But the chemist was also a physician, and one deeply interested in promoting the public health. A president of the Michigan State Board of Health, and a chairman of the American Medical Association's section on State Medicine and Public Hygiene, Kedzie was alert to the dangers of chronic intoxication from environmental poisons. Thus when his agricultural friends proudly exhibited calculations of the number of barrels of sprayed apples or heads of sprayed cabbage one would have to consume at a single sitting in order to poison himself, Kedzie corrected them. It could be true, he granted, that "any of the doses of arsenic . . . found in a pound of these fruits might be swallowed without endangering life by such single dose," but "it is the repeated doses, day by day . . . that might produce slow poisoning and the gradual undermining of the health without obvious cause."[51]

The criticisms of Hills and Kedzie were both aired in the

mid-1890s, nearly thirty years after the introduction of Paris green. Kedzie had been concerned with this aspect of agricultural chemistry from the beginning, and Hills was active in the agitation against domestic arsenic. Each might thus have been expected to condemn sprayed produce, yet even their opposition was far from adamant, being softened by "mays" and "mights." At a time when arsenical poisoning was a subject of intense medical debate, and when American farmers were generously applying ever-increasing quantities of arsenic to their fields and orchards, the American medical profession was curiously demure on the subject of spray residues and health.

The temptation is to credit the profession's reserve to ignorance, to suggest that because preventive medicine was a relatively new area of activity being explored by only a few men, and because these explorers were preoccupied with urban health problems, the rural practice of coating produce with arsenic was easily overlooked. Preventive medicine was indeed immature during the second half of the nineteenth century. It had begun its notable growth only in the 1840s in Britain, being nurtured by the realization that cholera and other infectious diseases could be prevented, and that prevention was preferable, in human and economic terms, to cure. Informed by statistical demonstrations that disease rates were proportional to levels of filth, the pioneers of preventive medicine adopted "sanitary reform" as their program and busied themselves cleaning out the dirtiest corners of the kingdom. This proved no easy task, for many corners, and not only those of Britain, were incredibly filthy. Cities had been growing since medieval times with a cavalier disregard for refuse disposal, but the recent increases in population density brought about by industrialization, and in America by immigration, had aggravated the filth nuisance nearly beyond human toleration. In 1865 a New York City neighborhood was described in terms that could have been accurately applied to most of the large cities of the Western hemisphere. "As a rule," a city inspector reported, "the streets are extremely dirty and offensive, and the gutters are obstructed with filth. The filth of the streets is composed of house-slops, refuse vegetables, decayed fruit, store and shop sweepings, ashes, dead animals, and even human excrements. These putrefying organic substances are ground together by the constantly passing vehicles. . . . When remaining moist or liquid in the form of 'slush,' they emit deleterious and very offensive exhalations. The reeking stench of the gutters, the street filth, and domestic garbage of this quarter of the city, constantly imperil the health of its inhabitants. It is a well-recognized cause of diarrhoeal diseases and fevers."[52]

As this passage suggests, the early sanitary reformers' understanding of the relation between dirt and disease was naive. It was an extension of the centuries-old view that decaying organic matter corrupts the surrounding atmosphere with disease-producing miasms. Whether one regards filth as the immediate cause of disease, however, or only as a vehicle for the transportation of disease from one person to another, the removal of filth results in the same benefit of improved health. Sanitary reform succeeded, if for the wrong reason, and in America as elsewhere during the late nineteenth century, "sanitary science" was the synonym for preventive medicine.

In this country, the advocates of the new science at first had to make up in enthusiasm what they lacked in numbers, for the American medical profession was slow in appreciating the value of preventive medicine. One of America's public health pioneers, Henry Ingersoll Bowditch (another Boston doctor), complained in his 1876 address on Public Hygiene and State Preventive Medicine that this subject was only seven years old in the United States. He selected 1869 as the birthdate of American preventive medicine because that was the year of organization of the first State Board of Health, a board, incidentally, of which Bowditch had been the first chairman. He had labored in that capacity ever since his appointment, and so had been understandably chagrined by the findings of a national sanitary survey he had conducted a few months before this address. The replies to the questionnaire he had mailed to prominent medical men in forty-eight states and territories demonstrated a "general lukewarmness of the profession in regard to public hygiene and preventive medicine," and revealed that "of our medical schools, only one or two . . . make any pretence at teaching either public or private hygiene."[53] The resolute Bowditch still managed an optimistic conclusion, promising that "in every civilized spot of God's earth . . . and to all coming time, human life will be lengthened, made more healthy, and consequently more truly happy, by the potent influence of STATE PREVENTIVE MEDICINE."[54]

Despite rhetorical stimulation, preventive medicine grew but slowly, and the number of medical activists occupied with the detection and elimination of preventable health hazards for some years remained so small in proportion to the number of detectable hazards that insecticide residues might have been disregarded. This possibility seems especially plausible in light of the fact that most of the attention of sanitary scientists was being directed to problems quite different from that of sprayed produce. The equation of dirt with disease had given preventive medicine a subterranean inclination, a preoccupation with providing a pure water supply and efficient sewage removal. As late as 1883, a member of Great Britain's Sanitary Institute expressed a hope that the "dangers to health above the ground such as are produced by poisons in various manufactured foods for domestic use, can only claim and will receive a like attention"[55] as had the subjects of water supply and sewage systems. In America, Bowditch had already discovered from his survey that only one-third of the state and territorial governments were concerned about the problem of adulterated food.[56] It might be added that even among those few sanitarians investigating food adulteration, manufactured and processed food items were bound to arouse more suspicion than nature's own fresh, presumably unalloyed, produce.

The presence of arsenic on fruit and vegetables might thus have escaped the notice of the medical profession. Doubtless a number of individual city doctors did remain unaware of the practice of treating produce with arsenic, but for the profession as a whole ignorance could not be the plea. References to Paris green as an insecticide were occasionally to be found in medical literature. Listings of available sources of arsenic for homicidal or suicidal purposes, in particular, usually included mention of the pigment used to destroy potato bugs, and some medical acquaintance with agricultural practice must be assumed. It was this uncertainty about the degree of danger from sprayed produce, more than ignorance of its existence, which kept doctors from seconding the warnings of Hills and Kedzie. The uncertainty about chronic intoxication from other sources of arsenic has been demonstrated. In the case of sprayed produce, the danger was even less distinct, for the reports of agricultural scientists who had investigated the matter indicated that the arsenic residue on produce sent to market was either extremely small or nonexistent. Medical critics of more probable arsenic hazards such as wallpapers had already been smeared with labels ranging from "faddist" to "crackpot," and were surely sensitive to the fact that to bring sprayed produce into their campaign would be to invite further ridicule. A very similar embarrassment had, in fact, been the fate of the health officers who had precipitated the fiasco of "the New York grape scare."

Readers of the *New York Times* of Friday, September 25, 1891, were startled by the headline "POISONED GRAPES ON SALE."[57] Reading further, they learned that two days earlier a concerned citizen had carried to the City Health Department some grapes he had recently purchased but feared to eat after discovering a peculiar green coating on them. The suspicious grapes were immediately analyzed by the Department's chemist, who reported the green matter to be copper sulfate, the active ingredient of Bordeaux mixture, a popular fungicide.† Knowing that New York growers sometimes mixed Paris green with the fungicide, the chemist raised the possibility that arsenic might also be present in "the green stuff" covering the grapes. He was, for some reason, unable to detect arsenic with certainty, but the presence of copper alone seemed sufficient reason for alarm. Copper, though potentially toxic, must be taken in much larger quantities than arsenic to produce either acute or chronic injury, and the amounts of copper on the New York grapes were quite small, less than the natural copper content of wheat, liver, chocolate, cucumbers, and a number of other common foods. The grapes were most likely harmless, but Health Department officials, surprised at finding copper artificially applied to grapes, overreacted. Copper sulfate residues, they feared, might "cause great suffering," and Department agents were at once dispatched to neighborhood groceries to confiscate any green-coated grapes they could find, as well as to wholesale houses to issue injunctions against selling any more such produce.

† Bordeaux mixture, a combination of copper sulfate, lime, and water, had been discovered by accident. According to the story associated with the fungicide, Bordeaux vintners had long suffered from the depredations of passersby who stole grapes from vines along the roadside. To combat this thievery, growers began to sprinkle their vines with a mixture of lime and copper salt (usually the sulfate), in order to give the grapes the appearance of being poisoned. At first, the only dividend of the practice was to discourage pillaging, but in 1882, when French vines were attacked by the downy mildew, a fungous disease imported from America, it was observed that vines growing along roadways were unaffected by the mildew. The only other difference between these and the vines hit so hard by disease was the coating of lime and copper on the healthy ones. Lime had already been tested against the mildew, and found to be inadequate, so copper salts were adopted as the basic ingredient in a number of fungicides soon developed. Bordeaux mixture proved the most popular, though, and had been adopted by farmers of other countries, including America, by the mid-1880s.

The Hudson Valley farmers who had shipped the grapes to market became alarmed for a different reason. Reading exaggerated journalistic accounts of the mass dumping of grapes in New York's harbor, and hearing more accurate reports that other eastern cities were considering grape boycotts, the farmers began to fear for their eastern grape trade. The U.S. Department of Agriculture was wired for help, and soon B. T. Galloway, Chief of the Division of Vegetable Pathology, was en route to New York for a conference with Health Department officials. After listening to the officials' side of the story, Dr. Galloway proceeded to show them data from experiments performed at the Department of Agriculture demonstrating that grapes sprayed in accordance with departmental recommendations for Bordeaux mixture never contained more than traces of copper; he explained that the amounts of copper found on the grapes seized by the Health Department were, though admittedly more than traces, still far too small to do any physiological damage; he finally took the health officials to task for their rash actions, which had allowed sensation-hungry journalists to "make a mountain out of a molehill."[58]

The journalists themselves quickly effected a very neat about-face, and were shortly characterizing the Health Department's actions as "reckless," and insinuating that the Department's chemist owed his appointment to considerations other than professional competence. Now defending the safety of Bordeaux-sprayed grapes, the *Times* hinted that grape dealers had a good case for suing the city for damages. Similar ideas had occurred to the growers of the Hudson Valley, who were threatening to take the city to court if the grape scare continued to interfere with trade. The Health Department backed down as gracefully as possible, acknowledging that only a small percentage of grapes had been found to be visibly contaminated with copper, and that even these were not dangerous. The Department still maintained that such grapes were "unwholesome," and requested that growers not ship green-coated grapes to market.

With this détente, public confidence in grapes was restored and grape growers spared from ruin. The financial loss inflicted upon farmers was not the only regrettable aspect of the New York episode, however, for the awkward defeat of the first public health agency to attempt to control the marketing of sprayed produce could only have served to make other health authorities more wary of intervening in similar situations. The New York Health Department lost a battle that was probably not worth fighting and, in the process, created a reluctance to engage in campaigns against the more serious threat of produce treated with arsenicals.

The forces of ignorance, toxicological uncertainty, and fear of professional embarrassment combined to hold medical censure of arsenical insecticides to a minimum during the nineteenth century. Admittedly, even had the medical profession been vigorous in its disapproval, Paris green would have continued in use. The necessities of insect control seemed to demand arsenicals; there was little legal machinery available to keep oversprayed produce off the market; and, as late as 1900, probably only a very small percentage of fruits and vegetables carried possibly harmful residues. Nevertheless, by their silence, medical scientists were passively cooperating with agriculturalists in the transformation of a potential hazard into an actual one.

*The use of poisons for the treatment of food
is a matter which calls for the closest
attention and the strictest control, where
it is not absolutely prohibited under severe
penalties. As in the case of food adulteration,
the public cannot be left to the tender
mercies of the interested or the ignorant.*

—editors, *British Medical Journal*, 1892

"Spray, O Spray" 3

THIS BIT of foreign advice, provoked by American agriculture's lack of restraint in the application of arsenical insecticides, might just as easily have been offered twenty-five years later. It would have been equally pertinent then, and the fact that at either date it would have been unsolicited is some indication of the nonchalance that allowed insecticide residues to achieve the proportions of a serious public health threat in the United States. The occasion for the early British warning that Americans were heading for trouble was a brief controversy initiated by England's *Horticultural Times*. In 1891, this popular journal announced that apples imported from the United States often were covered with a fine powder of arsenic, the residue of excessive spraying.[1] The first response of American fruit growers was to suggest that any powder adhering to the apples must be flour, since the fruit was often shipped in used flour barrels.[2] When this logical explanation was ignored by the British press, which continued its insinuations that American apples were deadly and the people who grew them daft, American pomologists sensed conspiracy and

68

suspected that English journalists were writing "in the interest of speculators for the purpose of injuring the sale of American apples in the English market."[3] The conspiracy was no doubt imagined, nor was there any certainty that the sinister powder was, in fact, arsenic in any form. Both sets of prophets, not surprisingly, proved wrong: no Englishmen were killed by foreign apples, and no Americans were impoverished by the loss of British markets. Nevertheless, the episode contained a lesson that, as the *British Medical Journal* observed, apples or any other produce might be poisoned by insecticides if no safeguards against such poisoning were adopted. Americans, of course, were no longer inclined to bow to British opinion, and this particular criticism went unheeded here. More than two decades later, a prominent federal entomologist could admit (though his statement carries no hint of reluctance) to a foreign inquirer that, "there exists, in this country, no regulation in force and no precaution is taken toward preventing accidents resulting from the use of arsenicals."[4]

The danger inherent in the farmer's complete freedom to use arsenicals was multiplied by the entomologist's encouragement to do so. The president of the American Association of Economic Entomologists presented the 1898 convention of his colleagues with the boast that "the entomologist of the present bears a very different relation to the public than he did a quarter of a century ago. New knowledge and new responsibilities have come to him; whereas then his opinion was presented and received as a gratuitous matter to be experimented with if convenient, his dictum now carries the force of authority, and often has the support of state and federal law."[5] These words were not mere convention rhetoric. Economic entomology's growth in public stature during the last quarter of the nineteenth century was indeed remarkable. In 1875 only a very few states hired entomologists to advise farmers on means of dealing with their insect enemies. By 1900 there were more than a score of State Entomologists vested with regulatory, as well as advisory, powers. In the decade of the 1890s nearly every state had enacted nursery inspection laws aimed at preventing the importation of destructive insects from other states, and had delegated the administration of these laws to official entomologists. Additional entomologists were employed by the state agricultural experiment stations and by the federal government, and given funds to develop and publicize improved methods of insect control. As the president of America's economic entomologists had claimed, farmers now listened when the entomologist spoke, and the message they received with regard to insecticides was not one of restraint.

The entomological ardor for spraying demonstrated in Chapter I never waned, though it at first failed to win over farmers at quite the anticipated rate. In spite of the popularity of arsenical insecticides, there were still farmers at the close of the century who preferred old-fashioned insect remedies, either because of a fear of poisoning themselves or their families, or because of a distaste for the messiness of spraying, or because of a false sense of economy—a belief that the new insecticides were too expensive. Entomologists could combat these reservations with arguments of their own: the cost of insecticides was less, over the long run, then the damages done by insects; the inconvenience of spraying was preferable to the laboriousness of non-chemical insect remedies; and the dangers of poisoning were much more remote than implied by the exaggerations of "a few ignorant alarmists."[6]

This last point of the safety of sprayed produce (i.e., sprayed produce that had been given sufficient weathering time before harvest) had been insisted upon for years, but it began to be urged with increasing frequency in the aftermath of the New York grape scare. At the first report of the Health Department's action, entomologists throughout the East had rallied to the defense of Bordeaux mixture, and though department officials were discredited in a matter of days, vindications of spraying continued to be issued for months.[7] "One would need to eat from one-half to one ton of these grapes—stems, skins and all—to obtain the least injurious effect,"[8] a Massachusetts entomologist concluded, while a New York colleague was even more confident, asserting that, "It is simply an absolute impossibility for a person to get enough copper from eating grapes to exert upon the health any injurious effect whatever."[9]

These denials of danger were directed at the specific question of copper residues only, yet pronouncements so sanguine might have been interpreted by some readers as implicit support for spraying generally. One of the publications inspired by the grape scare went considerably beyond implication, and exploited the opportunity to defend arsenic, in addition to copper, sprays. The U.S. Department of Agriculture's *Farmer's Bulletin 7* was the most widely distributed and authoritative work of its genre, and followed its convincing demonstration of the absolute safety of copper fungicides with a two-page defense of "Spraying From the Hygienic Standpoint." "Spraying," in this instance, referred to the application of arsenical mixtures, and was presented as an operation virtually free from danger. The head of the Agriculture Department's Bureau of Entomology, C. V. Riley himself, was called upon to assure readers "how utterly groundless are any fears of injury"[10] from sprayed produce.

Riley's sentiments had originally been expressed in an 1891 address to Boston's Lowell Institute and were at least questionable then. In later years, however, their frequent repetition would become dangerously misleading, for shortly after Riley's speech the average residues of arsenic on fruit and vegetables began to steadily increase. This increase was due, in part, to the growing numbers of farmers using arsenical insecticides. The opinion of the entomologist was gathering "the force of authority," and the few recalcitrant farmers who refused to be moved by persuasion could often be moved by coercion. By the end of the century, several states had passed into law requirements that all farmers either spray their crops or pay to have it done by the state.[11] Entomologists had urged such legislation for some time, on the grounds that farmers who did not spray imposed unfair financial hardships on their more conscientious neighbors. So long as just one farm in an area was left as an unsprayed refuge for insects, it was maintained, surrounding farmers would have to spray more frequently to prevent reinfestation of their fields. Doubts about the constitutionality of forcing farmers to spray hindered the wholesale adoption of the legislation, but even in states where farmers were left a choice, optimistic entomologists felt certain that "the nonspraying orchardist . . . will soon be worthy of a place in a dime museum."[12]

A more significant factor contributing to the increase of arsenical residues was the improvement of spray mixtures and equipment. The amount of insecticide residue on harvested produce is a function of not only the length of time since the last spraying and the prevalence of wind and rain during that period, but also of the natural adherence of the insecticide to the surface of the produce. One of the reasons for the supplanting of Paris green by lead arsenate was the latter's greater adherence to foliage. Increased adherence meant that fewer and lighter sprayings should be necessary, and this potential saving of time and money encouraged a search for specific adhesive agents to add to lead arsenate and make it even more resistant to weathering. A number of sticky substances, including molasses and fish oil, were tested for adhesive activity and, by the early 1900s, farmers usually added either casein or oil to their spray mixtures to achieve longer-lasting protection. These adhesives were also spreaders; they enabled the spray droplets to flow evenly over the fruit's surface and provide an arsenical armor free of chinks. The density of the armor, finally, could be increased by the use of improved pumps capable of drenching trees with a spray applied at pressures up to 200 pounds per square inch. With modern techniques such as these, any farmer could repeat the experience of a Washington entomologist who enthused that, "after the second spraying in one year in the Yakima Valley, three inches of rain fell in a few hours, yet where arsenate of lead was used there was no need of re-applying the spray."[13]

Entomologists were aware of the full implications of the recent advances in spray technology. They appreciated that more tenacious sprays carried not only a promise of increased protection against insects, but also a threat of heavier residues on harvested produce.[a] The conclusion that had followed from nineteenth-century studies of Paris green and London purple—that residues were quickly and almost completely removed by weathering—had to be reexamined for lead arsenate. The results of this reevaluation of the residue question are significant: they indicate that, as expected, spray residues were increasing during the early years of this century and, equally important, they demonstrate that, conscientious as they were in monitoring arsenic residues, entomologists remained ignorant of the true nature of the danger posed by these residues.

The rise of arsenic levels on produce can be traced through several articles printed in the agricultural literature during the first two decades of the century. In 1904, the Kentucky entomologist Garman reported that at least two children in his area had been made seriously ill by eating sprayed cabbage, though the fault, he concluded, lay not with the practice of spraying but rather with the negligence of farmers and consumers.[14] If farmers would refrain

a This threat was particularly ominous in the western states that were rising to prominence in the fruit industry early in this century. Three-inch rainfalls are a rarity in the Yakima Valley, as in the other fruit-producing areas of states like Washington, Oregon, and Colorado. The scarcity of rain in the western states made natural residue removal much less thorough there than in the orchards of wetter eastern states, while the fact that the Yakima downpour failed to dislodge freshly applied arsenate of lead suggests that weathering might have been insufficient to effect complete removal of the new spray mixtures in any locale. It will become clear that although western fruit usually carried the larger residues, the produce of the East was far from untainted.

from spraying close to harvest, and if consumers would make it a habit to remove the outer leaves of cabbage, no poisonings could occur. Garman himself had sprayed several plots of cabbage with different arsenical mixtures, a varying number of times, and had analyzed samples from each patch after removing the outer leaves. Little or no arsenic had been found on any of the samples, nor had Garman or any of his coworkers at the agricultural experiment station suffered any noticeable ill effects after eating produce from their cabbage patches.

The following year, a pair of California chemists suggested that the arsenic they had detected in so many of their state's wines was derived primarily from the sprays used to protect grape vines from insects.[15] Doctors Gibbs and James felt certain that the arsenical content of wines was too low to be injurious, but nevertheless thought it wise to avoid any further contamination lest wines become dangerous.

In 1906, the Illinois entomologist S. A. Forbes published the results of an experiment in which he had applied lead arsenate to three apple trees, picking and analyzing fruit from two of the trees the day after the last spraying, and taking samples from the third tree two months later.[16] Of the two trees whose fruit was examined immediately, one had been sprayed twice with a mixture of normal strength, and the other twice with a mixture containing four times as much arsenic, yet the residue found on apples from the latter tree was only slightly greater than that on fruit from the former. This was a comforting indication that overzealous farmers who applied insecticide in heavier doses than necessary were not significantly increasing the residues on their produce. It was countered, however, by the discovery that apples from the third tree still held, after two months, half the amount of arsenic that fruit from the other trees had held on the first day. Weathering was clearly much less efficient at removing lead arsenate than it was with the older insecticides; yet, Forbes concluded, "these facts are not at all alarming."[17] Two grains of arsenic was the recognized fatal dose, he observed, and therefore one would have to eat seven to eight pounds of apple *peelings* at one time in order to be poisoned by the fruit picked two months after spraying. Two ounces of peelings, a much more likely meal, he added, would represent only a mild *medicinal dose of arsenic.*

Forbes's calculations were accurate, but the reasoning accompanying them was decidedly less scrupulous. First, very little of the produce grown outside experiment stations went unsprayed the two months immediately before harvest, and one might expect that much of it carried residues greater than those found under Forbes's experimental conditions. Secondly, his argument that these residues were safe was based on the assumption that acute poisoning was the only hazard. Finally, his intended reassurance that the quantities of arsenic which apple-eaters of reasonable appetite might expect to consume were comparable to those often prescribed by physicians ignored the facts that there was a great deal of debate within the medical profession over the benefits of arsenical medication, and that people taking such medication were already ill and were being closely watched by their physicians for signs of arsenical intoxication. The wisdom of permitting healthy people with no medical supervision regularly to ingest medicinal doses of arsenic with their food should have been questioned, but Forbes was hardly alone in failing to do so.

In 1910, for example, W. P. Headden, entomologist from Colorado, announced that he had found arsenic on marketed fruit from seven states (from California to New York), and in urine samples taken from himself and two assistants after the three had eaten these apples.[b] None of

b W. P. Headden, *J. Econ. Ent.*, *3*, 32-36 (1910). Headden's chief reservation about arsenical sprays was that they were accumulating in the soil to levels sufficient to poison fruit trees. His articles condemning sprays for tree poisoning—Colorado AES, *Bulletin 131*, 1908; *J. Econ. Ent.*, 2, 239-245 (1909); *3*, 32-36 (1910); *Proc. Colorado Scientific Soc., 9,* the three had become disturbed, though, for a quick calculation had revealed they would need to eat ten pounds of fruit a day to receive the ordinary medicinal dosage. C. D. Woods, of the Maine Agricultural Experiment Station, calculated four years later that if one had eaten the apples sprayed in his experiments, "it would have taken about half a bushel a day to get the maximum medicinal dose."[18] Finally, Woods's neighbors at the New Hampshire experiment station reached a similar conclusion in 1917, supporting their defense of sprayed apples not only with the compulsory comparison of residue levels to medicinal doses, but also with the observation that the inhabitants of Styria commonly eat large amounts of arsenic, yet "appear to . . . live to old age."[19]

Directed by the entomologist W. C. O'Kane, the New Hampshire investigation was carried on over a period of five years (1912-1917) and included consideration of the magnitude of arsenical residues to be found on several fruits and vegetables, the factors affecting removal of these residues, and the potential danger of insecticides to livestock and people. In the case of apples, O'Kane and his colleagues learned that the method of picking had considerable effect on the amount of residue. Ordinary bare-hand picking was found to displace some residue, the wearing of cotton gloves effected still further removal, and the lowest residue levels of all were found on fruit that had been wiped with a cotton glove as it was picked. Weathering was also clearly efficient at removing arsenic, a comparison of the residue on apples picked three to five days after spraying with that remaining after seventy-five to ninety-one days revealing "a reduction in arsenical residues for the whole period of approximately 75 per cent due, apparently, to the effects of rain and weather."[20] In any event, no matter when or how picked, sprayed apples seemed entirely safe: even

345-360 (1910)—engaged him in a rather heated controversy over the supposed danger of arsenical soils (see E. D. Ball, *J. Econ. Ent.*, 2, 142-148 [1909]; 3, 187-197 [1910]).

the most heavily-coated apple selected three days after spraying carried only one-third of a medicinal dose.[21]

Strawberries and blackberries, having rougher surfaces than apples, were expected to hold larger residues, though even O'Kane must have been taken aback by the finding that strawberries from a plot picked two days after spraying contained 11.5 mg. of arsenic per berry! No calculations accompanied this announcement, but the alert reader could quickly compute that two generous medicinal doses of arsenic could be obtained from a single berry, and that a dozen of these berries might be fatal. These recently sprayed berries had not been exposed to rain, and residue figures for plots of berries given longer weathering periods were much lower. Under no conditions were the residues for strawberries low enough to satisfy O'Kane, though, and he felt obliged to caution that "strawberries that are fully formed or nearly so, should not be directly sprayed with arsenate of lead, unless they are to be thoroly [sic] scrubbed before using."[22]

Strawberries were the only sprayed produce that disturbed O'Kane, for he had experimental evidence that the quantities of arsenic found on other fruits and vegetables were harmlessly small. Desirous of more concrete data than comparisons of residues with doses of medicine, he and his colleagues had actually studied the effects of lead arsenate on the animal organism. Guinea pigs were fed daily portions of lead arsenate equivalent to the maximum residues O'Kane had found on sprayed apples. After maintaining the pigs on this "unit dose" of arsenic for some time and observing no signs of illness, O'Kane removed the arsenic from the animals' diet for a temporary "rest period," then reintroduced the arsenic at the level of two unit doses. Alternating rest periods with dosage increases, O'Kane was able to keep all his guinea pigs alive up to a level of eight unit doses. The first guinea pig succumbed at eight, the second at nine, and the last not until thirteen times the amount of arsenic found on apples had been added to his rations. Certainly guinea pigs were not people, and pigs might be less susceptible to the insults of arsenic than were human beings, but they were also only about one-hundredth the size of a man, and it seemed highly unlikely to O'Kane that a guinea pig could be one hundred times less susceptible. "Therefore, since guinea pigs can ingest daily, without serious poisoning, amounts of arsenate of lead equal to the average maximum found on six apples, we should not expect serious or fatal results to follow the daily ingestion of an equal amount by an adult person."[23]

The guinea-pig experiments were concerned solely with the evaluation of the threat of acute poisoning, but O'Kane's group was not unaware of the possibility of chronic injury. They cited a reference in a standard work on toxicology that attributed "what would appear to be serious poisoning" to an amount of arsenic, smaller than normal spray residues, which the victims had unknowingly swallowed daily, over an extended period, in their beer. Nevertheless, O'Kane and his coauthors were "inclined to suspect that other causes played a part in these cases."[24] Even if not due to unspecified "other causes," poisonings by such small amounts of arsenic seemed refutable by the example of the Styrian arsenic-eaters.

O'Kane's analysis of toxicological questions had not always been so carefree. At the beginning of his lengthy residue study, he had appreciated his lack of medical expertise, and had solicited an authoritative opinion, that of the University of Chicago's Anton J. Carlson.[c] Carlson was selected

[c] From this time until his death in 1956, "Ajax" Carlson (the name of endearment bestowed by his students) stood strongest among the medical opponents of lead arsenate and other hazardous pesticides. Carlson's career was an intellectual's fulfillment of the American dream: from the poor Swedish farm on which he was born in 1875, he emigrated to Chicago to work as a carpenter's assistant at age 16, obtained a master's degree in philosophy from Augustana College by 1899, and spent a year as a Lutheran minister in Montana before entering Stanford to study physiology. He earned his Ph.D. by 1902, and in 1904 accepted an appointment to the physiology department at the

not only because of his stature as a physiologist but, more specifically, for the investigations he had recently conducted with his colleague, Albert Woelfel, in determining the ability of human gastric juice to dissolve various inorganic salts. These studies were nearly unique for, by virtue of its location, gastric juice is not an easy fluid to examine closely. For centuries, the nature of digestion in the stomach was necessarily the subject of much speculation and little demonstration, until the mystery was finally solved in the 1820s by an American military doctor, William Beaumont. Beaumont exploited, when he could catch him, Alexis St. Martin, a young French-Canadian trapper afflicted with a gastric fistula, an open passage from the stomach to the abdominal surface resulting from a gunshot wound that Beaumont had been unable to repair completely. Dr. Beaumont more than made up for his lack of surgical finesse, however, by using Alexis' fistula to study the physiology of digestion, peeping through the hole into the stomach to observe digestion *in vivo*, and withdrawing gastric juice through the hole to study it *in vitro*. The collaboration between Beaumont and St. Martin was, understandably, not always a peaceful one, but it laid the foundation for the modern understanding of digestion, and became a classic in the history of physiology.[25]

By a remarkable repetition of history, a second sufferer from gastric fistula chanced to meet an American physiologist three-quarters of a century later, and to pick up where St. Martin had left off in offering his gastric juice to science. Fred Vlcek was a native of Bohemia who, at the age of six, had swallowed a solution of caustic alkali and severe-

---

University of Chicago. He never left the university, and his brilliant research work, combined with his popular teaching, eventually established Ajax Carlson as an institution unto himself. His biographer thinks it "probable that no man in America not engaged in clinical practice had so great an effect on medicine" (L. R. Dragstedt, *National Academy of Sciences, Biographical Memoirs, vol. 35*, New York, 1961, p. 1).

ly damaged his esophagus. Within five years his alimentary tube had completely closed, and Vlcek was having to take his meals through a gastric fistula provided for him by his doctors. The fistula was intended to serve only temporarily, until the esophagus could be surgically reopened, but when the patient overheard his doctors discussing their surgical plans, he became so frightened he fled the hospital in a coal wagon and eluded detection for fourteen years. Vlcek finally surfaced again, complete with fistula, in America in 1910, and shortly after became acquainted with Anton Carlson, who decided, like Beaumont before him, that Fred "should be made use of in the interest of physiology."[26]

One of the uses that Carlson made of his subject was as a source of gastric juice for studying the solubilities, in the human stomach, of several inorganic compounds, including salts of lead. Hence, Carlson was the logical man to answer O'Kane's questions about the solubility of lead arsenate. It was O'Kane's suspicion that lead arsenate is less soluble in gastric juice than is white arsenic (arsenic trioxide), that it is consequently less toxic (being absorbed into the bloodstream to a lesser degree), and that therefore comparisons of spray residues with lethal doses of white arsenic are even more demonstrative of the safety of spray residues than had been previously appreciated. Desiring expert confirmation of this suspicion, O'Kane requested Carlson and Woelfel to perform the necessary tests, and was soon informed that "arsenate of lead is . . . probably . . . less soluble [in gastric juice] than is . . . white arsenic."[27]

Delighted with these facts, O'Kane was clearly less than enchanted with the physiologists' interpretation. Carlson and Woelfel concluded their report to O'Kane with the proposal that "lead arsenate, used as an insecticide or spray on fruit trees, is sufficiently soluble in human gastric juices to cause lead and arsenic poisoning. Measures must, therefore, be taken to remove this spray from the fruit."[28] In a separate letter to O'Kane, Carlson had expressed his position in more detail: "Speaking as a physiologist interested in public health, I should say the question is not how much of the poison may be ingested without producing acute or obvious chronic symptoms, but how completely can man be safeguarded against even traces of the poison. There is no question in my mind that even in less than so-called toxic doses, lead and arsenic have deleterious effects on cell protoplasm, effects that are expressed in lowered resistance to disease, lessened efficiency, and shortening of life."[29]

Carlson's response was expert and unequivocal, and had been directly solicited by O'Kane. Nevertheless, by the time O'Kane published the results of his own experiments, two and one-half years later, he had developed a deeper faith in his own toxicological judgment and now rejected the physiologist's opinion. The conclusion that measures should be taken to remove spray residue from fruit, O'Kane pointed out, "should be understood as that of Drs. Carlson and Woelfel and not that of the writers of this bulletin. Our own conclusions . . . may be somewhat differently expressed."[30] They were, in fact, quite differently expressed: "Under ordinary conditions," O'Kane concluded, "no apples will reach the consumer carrying such amounts of arsenate of lead per fruit that a healthy human adult can eat enough at one time to cause fatal poisoning."[31]

O'Kane's paper has been discussed in some detail because it was the most thorough and best informed of any of the entomological studies of arsenical residues published prior to the 1920s. As such, it is the best illustration of the unequal tug-of-war swaying entomological opinion: people should be protected from poisons, but they must also be protected from insects, and while the insect menace was clear and present and could be evaluated in concrete terms of dollars and cents, even experts described the hazard of spray residues with nebulous phrases like "lowered resistance to disease" and "lessened efficiency." The imbalance between these opposed considerations easily tipped entomologists toward optimistic conclusions, and spray residues were generally dismissed as being considerably less dangerous than they actually were. In 1913 A. L. Quaintance, one of America's most respected economic entomologists, replied to the inquiry of a French agriculturalist that, in his opinion, the lowest harmful concentration of arsenic in food was one part per 7,000 (by weight), and since such levels were never observed on sprayed produce, "it seems permissible to conclude that all danger of injury . . . is entirely negligible."[32] Quaintance was apparently unaware that, by that date, many members of the medical profession had been adhering for a full ten years to a "world tolerance" for arsenic (a figure recognized internationally as the highest safe concentration of the poison) that was exactly one one-hundredth of the figure he had recommended as the lowest dangerous level of arsenic.

Your wine-tippling, dram-sipping fellows retreat,
But your Beer-drinking Briton can never be beat.
—British patriotic proverb

On the high seas and the battlefield, beer-drinking Britons no doubt were formidable, but in the enervating environment of the cities, ale could often prove an Englishman's undoing. During the late nineteenth century, particularly, alcoholic peripheral neuritis stemming from overindulgence in beer was one of the more frequent complaints encountered by doctors in English working-class districts. On several occasions, the incidence of alcoholic neuritis assumed epidemic proportions and, in the fall of 1900, still another epidemic seemed to be building.[33] Several physicians in Manchester observed a sharp rise in the number of patients complaining of neuritic symptoms, and though the phenomenon at first seemed due to "increased drinking associated with the recent elections and rejoicings in connexion [sic] with the war,"[34] closer study of the problem indicated some agent other than alcohol might be at fault. The epidemiology of the neuritis outbreak included two curious facts: all of the afflicted drank beer, but a number were only moderate or light drinkers; and, the large majority of all those affected were from the poorest classes, people who could afford to quench their thirsts with only "fourpenny" and "sixpenny" beer.

The possibility that the epidemic was due to a poison other than alcohol, and present only in cheap beer, was indicated by clinical data as well. Several doctors noticed that the disturbances of sensory and motor nerves were more severe than usual in alcoholic neuritis. Patients complained of swollen hands and feet, and of tingling and burning sensations in their extremities. In the suburb of Heywood, workers had begun to call the local brew "tenderfoot ale" even before the epidemic had been recognized, and a number had stopped drinking beer because of the pain it produced in their feet.[35] In addition to these signs of neuritis, beer drinkers were found to suffer from nausea, diarrhea, inflamed mucous membranes, and skin changes, including eruptions, growth of horny tissue, and darker pigmentation. The skin lesions, in particular, were suggestive of chronic arsenical poisoning, and when arsenic was found in the urine of many neuritis patients, the search for arsenically contaminated beer was on. It was not necessary to look far. Several medical investigators detected small amounts of arsenic in the cheaper beers of the Manchester area, and soon traced the poison to contaminated sulfuric acid used in the manufacture of glucose, a common substitute for malt in the poor man's beverage. The conclusion of one of these physicians—that arsenic "is contained in the beer in sufficient quantities to account for chronic arsenical poisoning"[36]—was not immediately concurred in by the profession at large, however. Doctors who had long relied on arsenical medication in the treatment of skin disease argued that "the drinkers of arsenical beer would have to swallow an ocean of the beverage"[37] to consume as much arsenic as their patients had taken without apparent ill effect. The beer's alcohol, others argued, intensified the action of what would otherwise have been harmless quantities of arsenic. In the discussion that followed, it was generally agreed that alcohol perhaps aggravated the effects of arsenic, but the fact that as soon as arsenical beer was removed from the market the incidence of alcoholic neuritis in Manchester dropped appreciably below its pre-1900 level indicated that arsenic was indeed the chief cause of the epidemic and that arsenical neuritis had been endemic to the area for some time.

Before its decline, the 1900 epidemic claimed an estimated seventy lives, caused more than six thousand illnesses, and generated such nationwide concern that a Royal Commission was appointed to investigate the dangers of arsenical contamination of foods. The Royal Commission Appointed to Inquire into Arsenical Poisoning from the Consumption of Beer and Other Articles of Food and Drink was chaired by the most venerable figure in British science, the physicist Baron Kelvin, and staffed by several of the most prominent medical specialists in the nation.

Appointed in February 1901, the Commission met regularly to examine physicians, chemists, and other experts capable of providing information regarding

(1) The amount of recent exception [sic] sickness and death attributable to poisoning by arsenic;

(2) Whether such exceptional sickness and death have been due to arsenic in beer or in other articles of food and drink, and if so,

    a) to what extent;

    b) by what ingredients or in what manner the arsenic was conveyed; and

    c) in what way any such ingredients became arsenicated, and

84

(3) If it be found that exceptional sickness and death have been due to arsenic in beer or in other articles of food or drink, by what safeguards the introduction of arsenic therein can be prevented.[38]

Included among the subjects discussed at the Commission's meetings was the danger of poisoning by spray residues. Arsenical sprays were a relatively recent addition to British agricultural practice, for although word of the efficacy of mineral poison insecticides had reached England soon after the American discovery, English farmers had been exceedingly reluctant to apply arsenic to their crops. Eleanor Ormerod, Consulting Entomologist to the Royal Agricultural Society, led a campaign to promote the use of Paris green, but her correspondence with American and Canadian entomologists reveals that she had great difficulty convincing Englishmen of the safety of the insecticide. Once a few began to use it, though, their successes won more converts, and by 1891 (the same year that British journalists accused Americans of exporting poisoned apples) Miss Ormerod could confide to a friend, "Surely it should be recorded of me, 'SHE INTRODUCED PARIS GREEN INTO ENGLAND.' "[39]

Not to belittle Miss Ormerod's triumph, the scale of use of arsenical insecticides in Great Britain was never comparable to that of America, though it was sufficiently great to attract the Royal Commission's attention. One of the commissioners asked Dr. Thomas Stevenson, vice-president of the Chemical Society and a lecturer at London's Guy's Hospital, about arsenical sprays, and was told that:

"in America they syringe the apples and the apple trees with arsenical compounds. . . . These, of course, are a source of conveying arsenic into the system . . .

"In poisonous quantity?

"In small quantities. It is difficult to say what is a poisonous quantity.

"Have you heard of arsenical illness traceable to such causes?

"Not in this country."[40]

The only related question asked by the Commission was put to a Mr. Berry, an agricultural expert, who replied that Paris green and similar substances were never applied to hops.[41] Brief though its examination of the residue hazard was, the Commission expressed a fear that uneducated and unrestrained farmers might apply arsenicals at too late a date, and in an appendix to their final report the commissioners reported an experiment in which gooseberries had been sprayed with Paris green. Berries picked a week after spraying (a week that included two heavy rainfalls) carried residues of 0.05 grain of arsenic per pound, and even fruit left to weather a full month still held 0.017 grain per pound. These residue figures are perhaps meaningless until examined in the light of the Commission's conclusions, presented some pages before this appendix. "Pending the establishment of official standards in respect of arsenic under the Sale of Food and Drugs Acts," the commissioners decided, "the evidence we have received fully justifies us in pronouncing certain quantities of arsenic in beer and in other foods as liable to be deleterious, and at the same time capable of exclusion, with comparative ease, by the careful manufacturer. In our view it would be entirely proper that penalties should be imposed under the Sale of Food and Drugs Acts upon any vendor of beer or any other liquid food, if that liquid is shown by an adequate test to contain 1/100th of a grain or more of arsenic in the gallon; and with regard to solid food—no matter whether it is habitually consumed in large or in small quantities, or whether it is taken by itself . . . or mixed with water or other substances . . .—if the substance is shown by an adequate test to contain 1/100th grain of arsenic or more in the pound."[d]

d Great Britain, *Royal Commission, etc. Final Report*, p. 50. The arsenic limits proposed by the Royal Commission, it might be noted,

A concentration of 0.01 grain of arsenic per pound of solid food was concluded by this commission of experts to be potentially harmful to public health. The Commission's recommendation, furthermore, was informally adopted by the British government[42] and the limit of 0.01 grain per pound or gallon soon came to be referred to as the "world tolerance" for arsenic as other nations recognized the importance and validity of the Royal Commission's work. Yet, the Commission's own experiments had shown that arsenic levels considerably in excess of this tolerance might be expected on fruit even a month after spraying, and soon American entomologists were reporting similarly high residues. Forbes, in 1906, found that apples picked the day after spraying carried 0.2303 grain of arsenic per pound (23 times the British tolerance). That farmers were free to, and perhaps did, spray fruit within days of harvesting needs no reiteration; the possibility that produce bearing twenty times the world tolerance of arsenic might be sold on the market was thus not so remote. Woods found approximately 0.006 grain of arsenic per pound on his Maine apples, a figure below the world tolerance, but hardly so far below as to justify Woods's contention that a half bushel of these apples could safely be consumed daily. The New Hampshire apples that O'Kane suggested could be eaten freely held as much as 0.008 grain of arsenic per *apple*, a concentration at least perilously close to the world tolerance, and probably considerably above it. Finally, A. L. Quaintance will be remembered for his assertion that one grain of arse-

---

were much more stringent for liquid than for solid foods. For instance, a limit of 0.01 grain per gallon of water is approximately 0.0012 grain per pound of water. The stricter tolerance on liquids was proposed because it was felt that arsenical beverages, especially beer, were more likely to be seriously contaminated than were solid foods. This difference in recommended limits, it will be seen, was later to be misinterpreted by entomologists who argued the commission had considered dissolved arsenic to be much more hazardous than solid arsenic, such as spray residues.

nic per pound was probably the lowest harmful level of spray residue.

The Manchester beer epidemic and the Royal Commission's report failed to draw much attention from agriculturalists. O'Kane made oblique reference to it (only to dismiss its evidence immediately), but otherwise the epidemic's relevance to the spray residue question was ignored. In 1904, J. K. Haywood mentioned the Manchester affair and warned of the "profound influences which even minute doses of arsenic exert upon health."[43] He listed green wallpapers and fabrics as sources of these "minute doses of arsenic," but ignored insecticides, even though he was writing in his official capacity as Chief of the Insecticide and Agricultural Water Laboratory, Bureau of Chemistry, United States Department of Agriculture!

American indifference toward the danger of arsenical produce did not pass unnoticed in other countries. The British had complained of American apples and apple-growers in the 1890s, and in the wake of the Manchester epidemic the French were to be even more dismayed by American agricultural practice. Interestingly, in France the employment of arsenic in agriculture antedated its use in America. Sometime in the late 1700s, French farmers had substituted white arsenic for lime as a treatment for grain blight, and had occasionally used it against insects as well, until an 1846 royal ordinance regulating the sale and use of poisons and narcotics. Article 10 of that ordinance forbade the use of arsenic as an insecticide, a prohibition that seems to have been closely observed at first, but that became a dead letter once the efficacy of arsenical insecticides was demonstrated in America. Most influential in restoring arsenicals to French agriculture was a M. Grosjean, an agricultural inspector who used his 1888 report to the Minister of Agriculture to tout Paris green and London purple. Too many farmers read and followed Grosjean's advice for the government to continue enforcing Article 10 of the 1846 law, and it was soon being ignored by farmer and official alike.[44]

In contrast to the French Government's silent agreement with farmers that arsenicals were too valuable to agriculture to be prohibited, the most prestigious institution of French medicine, the Académie de Médicine, subjected the insecticides to a rigorous, often noisy, examination. The most probing questioning of the safety of arsenicals came from Paul Cazeneuve, beginning with a paper read to his fellow academicians in 1908.[45] The most interesting aspect of Cazeneuve's paper, in light of the common American arguments in support of insecticides, was his emphasis on chronic, rather than acute, poisoning. He specifically cited the Manchester epidemic as evidence that even seemingly minute amounts of arsenic, amounts such as might be found on sprayed produce, were capable of inflicting injury if taken over long periods. He noted that similar epidemics had occurred much closer to home, presenting evidence to indicate that epidemics of neuritis in Paris in the late 1820s, and in Hyères in 1887, had been caused by arsenical contaminants. Citing several more recent cases of illness apparently caused by consumption of sprayed produce, Cazeneuve concluded with a proposal that all arsenical compounds be prohibited from agricultural use (i.e., that the 1846 ordinance be enforced in full).

Cazeneuve's position, though shared by many of his colleagues, looked like extremism to others, and the debate continued over several years. Cazeneuve's chief opponent in these discussions was a physician named Riche, who concentrated on the economic argument that the benefits of arsenicals outweighed the risks. Riche preferred to see Article 10 dropped from the 1846 law, and the first four articles, regulating the manufacture and sale of poisonous materials, given stricter enforcement. A sympathizing Dr. Weiss drew "applaudissements" for his observation that the Minister of Public Works had recently appointed a committee to investigate means of preventing accidents in the electrical industry, where accidents were considerably more serious and frequent than in agriculture. Did the opponents of arsenicals suppose, Weiss asked, that the public works commission would recommend forbidding the use of electricity?[46] The dangers of arsenicals, finally, could appear just as imaginary as their benefits were real. M. Linnossier dismissed any risk of chronic arsenicism from spray residues with his French version of a familiar argument. According to his calculations, Linnossier announced, one would have to drink five liters of wine made from heavily sprayed grapes to receive a dosage of arsenic equivalent to that found in a drop of Fowler's solution.[47]

A majority of the Académie, however, shared the sentiments of Cazeneuve, and of Armand Gautier, who insisted that "our principal objective, Messieurs, is the defense not of the prosperity of our vines and orchards but of that most precious treasure of all, the public health."[48] Their continued agitation to enforce or reform the ordinance of 1846 was a major factor behind the enactment of a 1916 law prohibiting the agricultural use of soluble arsenicals (insoluble arsenic compounds were to be permitted if conspicuously colored). This law seems to have had only slightly more success than its predecessor, for, as one of the Académie's cynics had remarked well before the passage of the 1916 law, there was no reason to expect farmers who had ignored the 1846 act to pay any more respect to a new one.[49]

Despite the failure to effectively control the use of arsenical insecticides, French physicians recognized their situation as superior to that of America, where no attempt at control had yet been made. Indeed, the American medical profession seemed wholly ignorant of the insecticide hazard, and French doctors often could not suppress their scorn. Gautier, commenting on the use of lead arsenate in dry powder form in the United States, noted that the farm workers who applied the powder "live in this plumbous atmosphere, breathe it, swallow it, and that, we are told, without danger, if one believes American documents."[50]

The entomologist authors of those American documents could not agree that their work was uncritical or irresponsible. The needs of farmers and the warnings of physicians had been placed in the balance, and the dangers of poisoning found too light. Sure of the rewards of scientific insect destruction, and doubtful that small amounts of arsenic could be injurious, entomologists gladly joined E. G. Packard in exhorting farmers to "Spray, O Spray":

Spray, farmers, spray with care,
Spray the apple, peach and pear;
Spray for scab, and spray for blight,
Spray, O spray, and do it right.

Spray the scale that's hiding there,
Give the insects all a share;
Let your fruit be smooth and bright,
Spray, O spray, and do it right.

Spray your grapes, spray them well,
Make first class what you've to sell,
The very best is none too good,
You can have it, if you would.

Spray your roses, for the slug,
Spray the fat potato bug;
Spray your cantaloupes, spray them thin,
You must fight if you would win.

Spray for blight, and spray for rot,
Take good care of what you've got;
Spray, farmers, spray with care,
Spray, O spray the buglets there.[51]

Packard's paean to poison exemplifies the attitude of the agricultural community toward arsenical insecticides during their first half century of use. Once converted to arsenicals, farmers devoted themselves to the cultivation of ever better gardens with a Panglossian optimism that assumed spraying could bring only good. The second half century of spraying was to begin with a renewed challenge of that assumption.

Part Two

Regulation

*The matter first came to the attention of our
department [the City Health Department of
Boston] quite accidentally through the
discovery, on August 18th by one of our
inspectors, of pears displayed for sale on
a fruit stand which were heavily spotted with
something which he did not know what it was
but which he deemed sufficient cause for
inspection. He took some of these pears to
the city laboratory of the State Health
Department [of Massachusetts] and examination
showed that they contained arsenic. From that
time on, we were active in gathering pears
throughout the city, which led to a general
inspection of pears and apples.*

—William C. Woodward, 1919

Regulatory Prelude   4

IN HIS grammatically involuted way, William C. Woodward, Health Commissioner of the City of Boston, thus
introduced federal public health officials to the existence
of a hazard from sprayed produce. It was a hazard that had
been suspected for some time by a few people with interests
in public health, but that had not been accorded official
recognition until the Boston inspector's confiscation of the
tainted pears. The inspector's Health Department had subsequently not only instituted a program of inspection of
all pears and apples in the city's markets, but had also re-
layed word of its action to the United States Department
of Agriculture's Bureau of Chemistry. The arsenical fruit

had been grown on the West Coast and shipped in interstate commerce, and hence was subject to federal regulation
under the 1906 Food and Drugs Act. That act had designated the Bureau of Chemistry as the federal agency responsible for initiating legal proceedings against adulterated
foods, and it was therefore the Bureau that was notified
first by Boston health officers. The Bureau's immediate response was to call a conference, held in Washington, D.C.,
on September 30, 1919, at which Dr. Woodward informed
an audience of Agriculture Department executives of the
common contamination of marketed fruit with excessive
residues of arsenic. From the date of this conference forward, spray residues were to attract increasing attention
from pure-food officials, and were eventually to become the
single most serious concern of these officials. To appreciate
the later efforts of health officers to minimize the danger of
arsenical produce, however, it is essential to survey the evolution of pure-food regulation in the decades immediately
preceding the 1919 conference. Precedents of regulatory
attitude and policy established during these years were to
largely dictate the federal government's handling of the
spray residue problem.

The adulteration of food is a deceit as old as urban
civilization.[1] In any society divided between food producers
and consumers, in which large numbers of people no longer
prepare their own bread and wine, the bakers and vintners
and other food vendors are constantly tempted to increase
their profits by debasing their products with cheaper ingredients. Several writers in antiquity complained of adulterated foods and beverages, yet the abuse evidently continued
and grew, for by the close of the Middle Ages ordinances prohibiting a variety of forms of adulteration had
been adopted by many European cities. Still, the nuisance
of adulteration did not become acute until the nineteenth
century, when the rapid urbanization accompanying the
Industrial Revolution created a situation in which growing numbers of city dwellers were made dependent on a

food supply shipped in from rural areas, much of it then canned or otherwise processed at a factory, and most of it finally sold by merchants who had had no hand in its preparation. The opportunities for adulteration were increased, and so were the temptations. The manufacturer's remoteness from the ultimate consumers of his product loosened the strictures of conscience, and he could sell to anonymous masses debased items he would never have offered a neighbor. This moral detachment also helped the manufacturer overlook the possibility that some of his adulterants might be poisons, and the nineteenth-century buyer truly had to beware lest the food he bought robbed his health as well as his pocketbook.

Adulteration was further encouraged by the progress of chemistry during the century. By the late 1800s, the chemical laboratory had yielded such a harvest of preservatives and artificial coloring and flavoring agents that the ingenious manufacturer could create an apparently wholesome, high-quality product out of cheap counterfeit ingredients. Yet the same chemistry exploited by the adulterator provided his undoing. The development of reliable quantitative and qualitative analytical techniques during the late eighteenth and early nineteenth centuries enabled the chemist to separate and identify the various ingredients of a food mixture, and thus to detect adulterant materials. Englishmen were particularly energetic in pushing the danger of food (and drug) adulteration before the public eye. The muckraking chemist Frederic Accum's *A Treatise on Adulterations of Foods and Culinary Poisons* (1820) created a market for frightening "death in the pot" literature, while the laboratory studies of Arthur Hill Hassall, published during the 1850s, gave such exposés sober substantiation.[2] The pure food and drug laws enacted throughout Europe during the second half of the century were the result of this publicity.

A very similar course was followed in the United States, though at a measurably slower pace.[3] By the close of the nineteenth century, most states had adopted statutory measures of some sort to combat food adulteration, and the federal government had outlawed the importation of adulterated food and drugs from foreign countries (a prohibition, incidentally, which was not strictly enforced). But the major traffic in impure foods—that of products transported in interstate commerce—was being allowed to continue unhindered. The individual states were powerless to act against such products, and the federal government, though endowed by the Constitution with the power to "regulate commerce among the several states," was reluctant to exercise it. Congress was still gingerly feeling its way into the area of governmental regulation of big business, the first large step in this direction, the Interstate Commerce Act, having been taken as recently as 1887. Federal regulation of the composition of food, furthermore, seemed to many so deep an intrusion into the affairs of everyday life as to represent an interference with the rights of the states, and even a threat to individual freedom. As one critic of the idea of federal legislation against impure food saw things, such regulation of the table menu was but a step away from a governmental attempt "to prescribe the table etiquette and dress."[4]

Partly as a result of this obstinate American independence, partly due to the lobbying efforts of certain food and drug manufacturers protecting vested interests, all nineteenth-century legislative attempts to prohibit adulterated foods from interstate commerce came to naught. Bills providing for such prohibition were frequently introduced into Congress after 1879, but, as the leading proponent of pure food legislation observed, "there seemed to be an understanding between the two Houses that when one passed a bill for the repression of food adulteration, the other would see that it suffered a lingering death."[5]

The grand total of these mistreated bills surpassed one hundred before one finally succeeded in passing both houses of Congress and becoming law in 1906. The Pure Food and

Drugs Act, as it was popularly known, was a victory of aroused public opinion over Congressional inertia, a victory engineered by a coalition of forces: by scandal-mongering journalists whose magazine exposures alarmed the public about the poisons in foods and drugs; by women's clubs that pressured legislators to insure a wholesome food supply for their constituents' families; by state food and drug officials who argued for the necessity of regulation on a national scale; and by many of the food manufacturers themselves, who hoped to purge their business of its less scrupulous operators.

But, above all, the 1906 law was a triumph for the indefatigable Harvey Washington Wiley, the man who, as Chief of the Agriculture Department's Division of Chemistry (changed to Bureau of Chemistry in 1901), had led the fight against impure foods for more than twenty years. Wiley was a Hoosier farm-boy by upbringing, a chemist and physician by education, and a man who combined the best elements of each heritage. His proficiency as a scientist earned him the directorship of the Division of Chemistry in 1883, and the hatred of dishonesty and fraud instilled in him in childhood moved Wiley to use the tools and power of his office to uncover the chemical sophistications of food adulterators.[a] His first annual report as division chief included a criticism of the adulteration of maple syrup with glucose, and an appeal for funding of chemical analyses of dairy products throughout the country.[6] Subsequent annual reports included a similar attention to adul-

[a] To be sure, the Division of Chemistry had displayed concern about adulteration before Wiley's arrival. Wiley's immediate predecessor at the head of the division, Peter Collier, had in fact studied various adulterated foods and urged that "Laws should be made and vigorously enforced making the adulteration of foods and medicines a criminal offense. Where life and health are at stake, no specious arguments should prevent the speedy punishment of those unscrupulous men who are willing, for the sake of gain, to endanger the health of unsuspecting purchasers" (quoted by C. W. Dunn, *Food, Drug, Cosmetic Law Quarterly, 1*, 309 [1946]).

teration problems, but these were merely supplementary to the comprehensive treatment of adulteration provided by Wiley's pet project during his first decade at the Division. *Bulletin 13* of the Division of Chemistry was entitled *Foods and Food Adulterants*, and dealt with the impurities commonly added to dairy products, spices and condiments, alcoholic beverages, lard, baking powders, sugar products, tea, coffee, cocoa, and canned vegetables. Published in several installments, beginning in 1887, Wiley's *Bulletin* revealed such practices as the dilution of whole milk and wine with water, of pepper with dirt, and of coffee with chicory and cereal. Aniline dyes were found in candies, and toxic metals in canned vegetables. Yet, disturbing as its disclosures were, *Bulletin 13* was not designed to alarm either the public or Congress. *Foods and Food Adulterants* was a rather technical volume, one of invaluable assistance to state chemists employed in food regulation, but too heavy for general public consumption.

To inform the layman about "the extent and character of food adulterations," a short treatise with that title was issued as *Bulletin 25* of the Division of Chemistry in 1890. Authored by the agricultural writer Alexander Wedderburn, the bulletin proclaimed that virtually every food product on the American market was adulterated to some extent. The majority of adulterants, Wedderburn maintained, were of the economic type, additives that cost the consumer money but did no injury to his body. Nevertheless, poisonous adulterants were not uncommon, and citations of cases such as that of the little girl killed by a copper-greened pickle provided poignant support for Wedderburn's plea for pure food legislation. The public was unfortunately not moved all the way to an active demand for anti-adulteration laws until 1906, when the descriptions of conditions in Chicago's meat-packing plants contained in Upton Sinclair's socialist novel *The Jungle* generated a national wave of revulsion that swept into the halls of Congress. Until this time, however, Bulletins 13 and 25, as well as subsequent studies performed at the Division of Chemistry, all directed and publicized by Wiley, had served to maintain food adulteration as a matter of continuing public concern. The 1906 law was precipitated by Upton Sinclair,[b] but the way had been cleared by Harvey Wiley, and it was Wiley who was deservedly given popular recognition as "father of the pure food law."

Wiley's law was "an act for preventing the manufacture, sale, or transportation of adulterated or misbranded or poisonous or deleterious foods, drugs, medicines, and liquors, and for regulating traffic therein. . . ." The act specified "that the examinations of specimens of foods and drugs shall be made in the Bureau of Chemistry . . . and if it shall appear from any such examination that any of such specimens is adulterated or misbranded within the meaning of this act, the Secretary of Agriculture shall cause notice thereof to be given to the party from whom such sample was obtained. Any party so notified shall be given an opportunity to be heard . . . and if it appears that any of the provisions of this act have been violated by such party, then the Secretary of Agriculture shall at once certify the facts to the proper United States district attorney . . . it shall be the duty of each district attorney to whom the Secretary of Agriculture shall report any violation of this act, or to whom any health or food or drug officer or agent of any State, Territory, or the District of Columbia shall present satisfactory evidence of any such violation, to cause appropriate proceedings to be commenced and prosecuted in the proper courts of the United States, without delay, for the enforcement of the penalties as in such cases herein provided."

The judicial mechanisms available to district attorneys

---

[b] This most significant result of Sinclair's book was quite unexpected. *The Jungle* was intended to foment a much greater social revolution, to do for the white slaves of capitalism what *Uncle Tom's Cabin* had done for black slaves. Sinclair had set his sights too high: as was later observed, he had aimed at the public's heart, but hit its stomach.

prosecuting violations of the act were two: criminal proceedings against individuals could lead to sentences of fine and/or imprisonment; libel for condemnation proceedings against adulterated or misbranded products could result in the products' being disposed of by sale or destruction (though the original owner of any such product could repossess all samples seized by the government by paying the costs of the libel proceedings and swearing not to remarket the product in its adulterated or misbranded form).

The Food and Drugs Act is remarkably free of confusing legal phraseology, and seems quite clear and straightforward on first reading. Bureau of Chemistry personnel did not need much enforcement experience, however, to discover a number of inadequacies in the language of the law. The act strove to protect the consumer from being cheated by defining food as adulterated "if any substance has been mixed and packed with it so as to reduce or lower or injuriously affect its quality or strength," and as misbranded if its "package or label . . . shall bear any statement, design, or device regarding such article, or the ingredients or substances contained therein which shall be false or misleading in any particular." But the law did not provide guidelines or standards for determining what was a food's normal "quality or strength," or for defining "ingredients and substances" so as to allow clear distinctions between imitation and genuine articles. These areas of ambiguity were left for the Bureau of Chemistry to clarify, and Wiley and his colleagues soon came to appreciate that such seemingly simple questions as "What is whiskey?" or "What is a sardine?" could excite near interminable controversy.[7]

An even thornier problem bequeathed by the law to the Bureau was the detection of foods harmful to the consumer's health. The Food and Drugs Act had forthrightly provided that food must be considered adulterated "if it contain any added poisonous or other added deleterious ingredient which may render such article injurious to health." But the questions of which particular ingredients should be designated "poisonous" or "deleterious" proved a Pandora's box once opened. At the head of the host of perplexities that emerged was the difficulty, encountered before in the case of arsenical wallpapers, of detecting chronic injury and relating it to the ingestion of a specific adulterant. The large majority of food adulterants with which the Bureau would have to deal, Wiley realized, were substances present in too small amounts to produce acute illness; when deleterious, they were chronic poisons. Such adulterants were typified by a group of food preservatives that had recently come into general use and that was a special object of Wiley's scrutiny. His investigation of the physiological effects of these preservatives is a particularly instructive illustration of the deterrents to successful identification of an adulterant as poisonous, and prefigures many of the issues that later arose with spray residue regulation.

The development of national markets in the food industry during the second half of the nineteenth century seemed to many food manufacturers to necessitate the addition of certain chemicals to their products to protect them from spoilage during shipment to distant points. Among the more popular of these preservatives were benzoic, boric, salicylic and sulfurous acids, items necessary in only very small quantities presumably entirely safe for consumers. Wiley, however, was skeptical of this assumption, both because it conflicted with his scientific experience and, perhaps more importantly, because it offended his moral sensitivity. Wiley abhorred deceit, and was exasperated by the new preservatives that, unlike the salt, vinegar, and spices traditionally used to preserve food, could not be detected by taste or odor when mixed with food.[8] The consumer of modern preserved foods was thus being fed substances of which he was unaware and, to make the abuse worse, these substances were artificial chemicals. There was no doubt in Wiley's mind that synthetic chemicals were somehow less wholesome than natural products, if only because the Lord was a better chemist than man. "Of everything made by

man," Wiley once announced to a Congressional committee, "almost nothing has the hygienic value of that made
by nature."[9]

Wiley was still too much the scientist to rest his case
against preservatives entirely on *a priori* suspicion. He recognized the need for experimental data to substantiate his
opinion, and attempted to obtain it with his famous Poison
Squad investigations.[10] The Poison Squad was a group of
young, originally healthy male civil service employees who,
in the fall of 1902, began to answer Wiley's call for volunteers to undergo a feeding program to determine the effects
of various food preservatives on digestion and health. The
volunteers who were accepted were required to take all
their meals during the period of the study at Wiley's
"hygienic table" in the basement of the Bureau of Chemistry building. The fare at the table consisted of preservative-
free food during a "fore-period" of about ten days, during
which the volunteers' normal metabolism was studied, followed by meals mixed with the preservative under investigation during a "preservative period" lasting two to three
weeks. Serious injury to volunteers was carefully guarded
against, but the potential for danger could not be hidden
from newspaper reporters in search of exciting copy. Soon
the diners at the hygienic table were being celebrated na-
tionally as the heroic "poison squad":

> *O we're the merriest herd of hulks that ever the world
>   has seen;*
> *We don't shy off from your rough on rats or even from
>   Paris green:*
> *We're on the hunt for a toxic dope that's certain to
>   kill, sans fail,*
> *But 'tis a tricky, elusive thing and knows we are on
>   its trail;*
> *For all the things that could kill we've downed in many
>   a gruesome wad,*
> *And still we're gaining a pound a day, for we are the
>   Pizen Squad.*

This "Song of the Pizen Squad"[11] included Paris green among the imagined foods of the volunteers, it might be noted, simply because the pigment was a notorious poison for suicides, not because it was generally appreciated to be present on sprayed produce.

Wiley took such journalistic exaggerations in stride, and was not at all dismayed that his "poison-squad laboratory became the most highly advertised boarding-house in the world."[12] He did, however, appreciate that the use of human volunteers for his study was attended by certain difficulties. The squad members could not be confined in a controlled environment as animals could be, though Wiley attempted to regulate their behavior away from the table as far as feasible. Being aware of the nature of the food they were given, the volunteers were subject to psychosomatic disturbances that would not affect guinea pigs. Finally, their internal organs could not be examined for signs of the chronic lesions Wiley expected to be produced. In spite of these drawbacks, Wiley preferred people over animals for his tests because he suspected data obtained from lower animals might not be recognized as applicable to human beings. The chief chemist was thus compelled to base his study of the physiological effects of preservatives on data obtained from regular physical examinations of each volunteer, and on painstaking laboratory analyses of the physical and chemical composition of each man's urine and stools. The Poison Squad experiments were continued for five years, and included investigations of five commonly used preservatives: boric acid and borax, salicylic acid and salicylates, sulphurous acid and sulphites, benzoic acid and benzoates, and formaldehyde. All were concluded by Wiley to be injurious in the amounts normally ingested, with the injuries ranging from "malaise" to loss of weight, digestive disturbances, and headache.

Wiley's conclusions did not, of course, pass unchallenged. Food manufacturers who employed preservatives naturally questioned the assertion that their products were thereby made dangerous. When, after passage of the Food and Drugs Act, it appeared that Wiley intended to regard preservatives as adulterants, some manufacturers even called on President Theodore Roosevelt and urged him to place the determination of poisonous and deleterious ingredients in some hands other than Wiley's. Roosevelt offered Wiley a hearing, and at a January 1908 White House confrontation between Wiley and representatives of the food industry, the chief chemist characteristically seized the initiative and was pressing toward victory when he made the most unfortunate gaffe of his career. As Wiley often recounted later with martyr-like pleasure, he had just convinced Roosevelt that benzoic acid was a harmful adulterant when one of the manufacturers brought up the subject of saccharin, an artificial sweetener used in canned corn. His wrath aroused, Wiley neglected to wait for a presidential response, but immediately denounced saccharin as "totally devoid of food value and extremely injurious to health."[18] At this precipitate declaration, "the President changed from Dr. Jekyll to Mr. Hyde, and said: 'You tell me that saccharin is injurious to health?' I said, 'Yes, Mr. President, I do tell you that.' He replied, 'Dr. Rixey gives it to me every day. . . . Anybody who says saccharin is injurious to health is an idiot.' This remark of the President broke up the meeting. Had he only extended his royal Excalibur, I should have arisen as Sir Idiot."

To add injury to insult, Roosevelt acceded to the requests of industry and the very next day appointed the Referee Board of Consulting Scientific Experts. Designed as an independent body to pass judgment on the safety or injuriousness of common food additives, the Referee Board was composed of five of the nation's most prominent scientific researchers, including the Johns Hopkins chemist Ira Remsen as chairman.[e] The first task assigned the Referee, or Remsen, Board was an investigation of the physiological

[e] Saccharin, more than coincidentally, had been discovered in Remsen's laboratory.

effects of sodium benzoate, a preservative already concluded by Wiley to be harmful. The Remsen Board study differed from Wiley's earlier work in several respects.[14] The experiment was carried out on a new poison squad recruited from healthy medical students, the preservative was administered over a greater dosage range and was more thoroughly mixed with the food, and the volunteers were kept ignorant of the transition from "fore-period" to "preservative period" to minimize psychosomatic reactions. The most significant difference between the Remsen and Wiley investigations was in the conclusions reached. Sodium benzoate, the Board announced in 1909, was not injurious to health when taken in the amounts ordinarily used to preserve foods.

The Remsen Board decision, not surprisingly, touched off a heated exchange between the Board's supporters and the Wiley-led anti-benzoate forces, each side charging the other with prejudice, careless research, and unsound interpretation of data. In retrospect, it appears that both sides were guilty, though Wiley perhaps more than the Referee Board, since the passage of time has lent support to the conclusion that sodium benzoate is not seriously harmful in small amounts. The Remsen-Wiley disagreement is nevertheless relevant to the subject of spray residues, for it exemplifies the difficulty that well-qualified scientists can experience in trying to reach agreement on the injuriousness of a substance. The source of disagreement is the fact that a substance's chronic toxicity, or lack of it, is not very amenable to scientific demonstration. Wiley admitted as much in a comment on the effects of formaldehyde used as a preservative. "It is evident," he maintained, "that the system is able for some time to control the development of conditions which later become pronounced, and that no ill effects are produced prior to that time is not probable."[15] Since the early effects of chronic poisoning are internal injuries with no accompanying external manifestations, their existence can be a matter only of conjectural probability and not of actual demonstration. To Wiley it seemed obvi-

ous that any abnormal constituent of food, such as a preservative, must interfere with the natural, normal activity
of the body's organs, and "any change in the food which
adds a burden to any of the organs, or any change which
diminishes their normal functional activity, must be
hurtful."[16] Such common-sense reasoning did not sound
compelling to all, and at the opposite pole from Wiley
might be found scientists requiring clear clinical signs of
physiological damage before judging a substance harmful.
This is assuredly a case of locking the house after the bur-
glar has entered, but one might argue that such is the only
certain way of catching the guilty party. In practical terms,
the disagreement condenses to the question of whether a
suspected adulterant should be presumed innocent until
proved guilty, or guilty until proved innocent. One approach imposes a risk on the public, the other a hardship
on business. The compromise effort to be fair to consumers
and producers alike is a near impossible task that, as will
become apparent in the Bureau of Chemistry's handling of
spray residues, is rarely appreciated by either side.

The Bureau's enforcement problems did not by any
means stem entirely from this difficulty of getting experts
to agree on dangerous adulterants. The Remsen Board investigated other preservatives in addition to sodium benzoate, and in some instances agreed with Wiley that the substances in question could be injurious. But even on these
happy occasions, industrial apologists could muster other
arguments to temporarily confound Bureau officials. One
of those most frequently enlisted in the pro-preservative
cause, and one particularly annoying to Wiley, was the observation that so-called adulterants often occur naturally in
a variety of foods. Wiley was repeatedly confronted with
the statement that cranberries in their pristine, unprocessed
condition already contain more benzoic acid than any preserved foods. This was true, Wiley readily admitted, and
not only cranberries but many other foods naturally con-

tain small amounts of poisonous compounds. But then, after assuring his opponents that he was "not criticizing the Creator at all for putting" poisons in cranberries and other foods, Wiley suggested that the presence of these substances in natural foods "instead of being a warrant for using more of them, points to the necessity of reducing their quantity to the minimal amount possible."[17] Wiley's logic hardly laid the "natural poisons" argument to rest. It was to be frequently revived in later years to rebut critics of adulterants other than benzoic acid, including arsenical spray residues.

Yet even admitting the folly of adding more poison to the food supply because some existed there naturally, might not a possibly injurious substance be permitted if its addition was necessary to mitigate a greater evil? It was the contention of catsup manufacturers, for instance, that poisoning from spoiled tomatoes was a more serious health hazard than benzoic acid poisoning, and that food treated with the preservative was therefore less dangerous than food not given this chemical protection. Wiley assented to the logic of this argument,[18] but suspected that many of the greater evils that chemical preservatives were being used to overcome were not necessary evils. The problem of catsup spoilage, he maintained, could be avoided by cleaner and more careful processing, and without the use of preservatives. Chemical additives must not be permitted, Wiley felt, where their function was essentially to camouflage sloppy manufacturing practices or merely to improve the product's appearance. Wiley aroused the ire of the fruit industry very early in his reign as chief enforcer of the Food and Drugs Act by his criticisms of the practice of bleaching and preserving dried fruit with sulfurous acid. "The question," Wiley noted, "is one worthy of very careful consideration— whether for the sake of preserving a light color and securing immunity from mould and decay it is advisable to introduce into a food product any quantity whatever of a substance injurious to health. The answer to this question seems almost unavoidable, and it is, and should be, negative."[19]

Even Wiley appreciated that a broad application of such a philosophy was unrealistic, that for many adulterants it was neither technically feasible nor toxicologically justified to prohibit the inclusion of "any quantity whatever" of a deleterious ingredient. Wiley was clearly distrustful of any, even the smallest, amount of a poisonous substance,[20] but toxicological opinion generally recognized the existence of a threshold value for a substance, a level at which it presumably could begin to affect normal physiological functioning, and below which it was innocuous. Accordingly, Bureau of Chemistry enforcement procedure included observance of a "tolerance" policy: potentially harmful additives were permitted in foods up to certain quantitative limits, and legal action was instituted only against products in which tolerance limits were exceeded. The tolerances observed were sometimes precise, such as the 350 mg. of sulfur dioxide per kg. of fruit originally recommended as a guide for regulating dried fruit. At other times, the tolerance could be impossibly vague, as with the "excessive amount" adopted as the limit for copper in green vegetables.[21]

The level at which any tolerance was set was determined by several considerations, the first being toxicological data. There was considerable epidemiological and experimental data available to indicate the dangerous amounts of older common poisons like copper and arsenic. Tolerance decisions for the newer additives had to be based on the results of the Poison Squad and the Remsen Board studies. Rarely, as indicated previously, was there a consensus on exactly what quantity of any of these substances constituted a minimum toxic dose. The room for disagreement was further enlarged in the case of the preservatives by the precaution taken by both Wiley and the Remsen scientists to use only healthy young volunteers on their poison squads. The young and robust, unfortunately, are a minority in society, and there are many people either very young, aged, or constitutionally weak whose susceptibilities to adulterant poisons might be expected to be significantly higher than that of the average experimental volunteer. Their tolerances could not be determined by poison squad data, and provision of adequate protection for all consumers thus depended to a marked degree on guesswork. Each and every adulterant had a dosage spectrum clouded by a hazy region between zero and the dose capable of demonstrable injury to the most vigorous, and finding in this region the point at which a substance ceased to be safe for all and became harmful for a few was a toxicologist's nightmare.

Nor was the location of a tolerance entirely within the province of toxicology. Even if a safe tolerance could be determined precisely, it would have no regulatory value unless food manufacturers were able to meet it. It may be recalled that the Royal Commission on Arsenical Poisoning settled on the tolerance it did, both because larger amounts of arsenic seemed "liable to be deleterious," and because they were "capable of exclusion, with relative ease, by the careful manufacturer." The cooperation of the governed is essential to the success of any law, and the Bureau of Chemistry, like the British Royal Commission, appreciated that the ability of a manufacturer economically to remove an adulterant from his product had to be considered in setting a tolerance on the adulterant. This concession to manufacturing interests was necessarily made at the expense of consumer safety. It was possible that the adulterant level attainable by the manufacturer would be lower than the tolerance a toxicologist might recommend, and if the lower figure were adopted for enforcement purposes, the public would be given an added measure of safety. Food technology was not always equal to the task, however, and when industry found a toxicologically advisable tolerance too low for immediate attainment, Bureau officials had little choice but to increase the tolerance. To achieve a workable policy it might be necessary for the public to consume possibly harmful quantities of adulterants for the period needed by industry to equip itself to satisfy safe tolerance standards. The Bureau's operating assumption had to be that this period of industrial adjustment would be too temporary for any consumers actually to suffer chronic injury, yet these hopes were not always realized and just how long these periods of adjustment could run would be made distressingly clear by the Bureau's experience with spray residue regulation.

The law itself often forced the Bureau of Chemistry to liberalize its tolerances, for the Food and Drugs Act provided that violations detected by Bureau inspectors must be tried in the federal courts. A jury of laymen ultimately decided if an indicted product had been adulterated within the meaning of the act. To a point, this provision was not objectionable: the decision as to whether or not a potentially deleterious ingredient had been added by the manufacturer, if not always quickly reached, was at least within the realm of competency of the average man. But it was not enough for the Bureau to demonstrate that an adulterant was present and that its tolerance had been exceeded, for the tolerances observed by Bureau officials had not been sanctioned by the Food and Drugs Act. They had no legal status, but were simply administrative guidelines for Bureau inspectors, to direct them when to seize a product as dangerous. The successful defense of a seizure action in court thus hinged on the Bureau's ability to vindicate the administrative tolerance in question. The evaluation of the validity of a tolerance was a matter demanding a certain amount of scientific expertise, not an appropriate task for uninformed laymen. Yet legally the evaluation had to be performed by a jury, and the Bureau was repeatedly frustrated in its attempts to convince jurors that such small amounts of a substance as seemed to them insignificant could in fact be harmful. Tolerances low enough to satisfy the toxicologist were often too low to frighten a jury, and to make its seizure actions defensible in court, the Bureau frequently had to set its tolerances somewhat above the figure considered safe. If this were not done, it was feared, prosecutions based on low tolerances would be rejected by juries, and manufacturers would be encouraged to adulterate products more freely. Liberal tolerances seemed the lesser of the evils.[d]

The Bureau of Chemistry's enforcement strategy, finally, was molded by political pressures. Manufacturers whose products stood to be classed as adulterated predictably opposed any strict interpretation and enforcement of the Food and Drugs Act. The appeal of some of these to President Roosevelt has been noted, though presidential intervention in food regulation was exceptional. More common was the

[d] It might be added that juries could present the Bureau with other frustrations. Violations of the act were tried in the district court of the area where the violation occurred, and jurors could find it difficult to punish one of their local manufacturers. A district judge from New York once sat on a case brought by the government against a carload of canned cherries in which the prosecution's evidence was so conclusive the judge assumed the jury "would shortly return . . . a verdict in favor of the government." Instead, the jury deliberated for hours before finding for the defendant, and accompanied its verdict with an apology. The foreman, a local farmer, explained, "We, the members of the jury, recommend to the stockholders of the canning company that they call a meeting at once and put new officers and managers in control of the plant. It will be very harmful to the fruit industry of Western New York if cherries such as these are shipped over the country. No such pack should ever have been put out, but, inasmuch as the stock of the cannery is owned by the local fruitgrowers, and since they did not know of the condition of the pack, and since its seizure would inflict losses upon them that they can't afford, we decided to give our verdict in favor of the cherries" (quoted by J. C. Knox, *Food, Drug Cosmetic Law Quarterly, 1,* 433 [1946]). On another occasion, government lawyers were confident of winning an oyster adulteration case tried in North Carolina. A local attorney warned them, "Gentlemen, you probably have heard of a little unpleasantness we had down here some sixty years ago known as the Civil War. These jurors are not going to convict their neighbors on the say-so of a bunch of Yankees from Washington." The quickly returned verdict was "not guilty" (P. Dunbar, *Food, Drug Cosmetic Law Journal, 14,* 114 [1959]).

advocacy of a particular industry's cause by congressmen representing states in which that industry flourished.

But most distressing, perhaps, was resistance to vigorous enforcement from within the Department of Agriculture itself. The numerous and varied functions of big government inevitably produce situations in which conscientious civil servants can find themselves working at cross purposes, and the Bureau of Chemistry was in one of these situations throughout its history. The Bureau was the logical governmental agency to be assigned the responsibility of detecting food adulterations: Bureau Chief Wiley had "fathered" the 1906 Act, and in the process the chemical analysis of foods had become a specialty of the Bureau's scientific staff. Just as logically, the Bureau was housed within the Department of Agriculture, the science of chemistry having been recognized since the mid-nineteenth century as essential to agricultural progress. Still, despite the power of logic, a serious conflict of interest developed between Bureau and Department after 1906. The Department of Agriculture, after all, had as its *raison d'etre* the protection and advancement of the prosperity of America's farmers, the same farmers who, as the ultimate producers of the country's food, were necessarily affected by the enactment and enforcement of pure food legislation. The 1906 Act provided that the Secretary of Agriculture report violations detected by the Bureau of Chemistry to the proper U.S. district attorney for prosecution, and one can readily appreciate the dilemma of a secretary informed by his chief chemist that a product such as dried fruit preserved and bleached with sulfurous acid was dangerously adulterated. The secretary's duty to protect the public from unwholesome food contended with his duty to promote the interests of thousands of Western fruit-growers who considered sulfurous acid vital to their livelihood. The resolution of such a conflict could hardly be expected to wholly agree with the Bureau of Chemistry's position. When action against sulfurous acid-treated fruit was first recommended by Wiley in 1907, in fact, Secretary of Agriculture James Wilson postponed confiscation of adulterated fruit pending further investigation of the problem, and promised fruit-growers that in the meantime they could "go on as you used to go on and I will not take any action to seize your goods or let them be seized, or take any case into court."[22] The responsibilities of their position made subsequent Secretaries of Agriculture subject to similar differences of opinion with the Bureau of Chemistry, and eventually this unfortunate administrative arrangement of the Bureau within the Department of Agriculture was to exert its hampering influence on the regulation of spray residues.

In view of the political, legal, and scientific realities of the Bureau's situation, it is not surprising that enforcement of the Food and Drugs Act was less than rigorous. Even when violations were successfully prosecuted, sentencing judges tended to be lenient. The law allowed a maximum fine of $200 as punishment for a first offense, and maximum penalties of a $300 fine and/or one year's imprisonment for subsequent offenses. The average fine levied for the first forty convictions under the act, however, was sixty-seven dollars, and fines were frequently five dollars or less, some being as low as one cent! Several spokesmen for drug manufacturers commented that when caught violating the act it was cheaper to enter a plea of guilty and pay a fine than to bother with the expense and trouble of a court trial.[23]

To be sure, the maximum penalties for food and drug violations were not particularly severe, but at the time of passage of the act harsher penalties had seemed unnecessary. Wiley himself at first believed: "It would be the rarest thing in the world that any suit would ever be brought under this law. Just the moment this law is passed the business men of this country, practically all of whom are law-abiding citizens, are not going into the jaws of the law. None of them that knows anything about the law will want to do that. As soon as the law is understood and fully appreciated, they will all conform to it."[24] Most of what violations there might be, Wiley felt sure, would be the result of "ignorance or pardonable carelessness," and in these instances it would suffice "to admonish the culprit and give him a chance—like the bad boy he is spanked by his fond father and given another chance to be good."[25]

The analogy of a governmental father fondly supervising the activities of his industrial children could suggest other ways of keeping these children out of trouble. Many children find the moral lecture more instructive and reforming (and unpleasant!) than corporal punishment, and certainly as the Bureau's enforcement policy evolved, there emerged an inclination to spare the rod until educational efforts had clearly failed. By the mid-1920s, the period during which spray residues were becoming a subject of major concern to the Bureau, assistant chief Paul Dunbar could explain that if an infraction "is one which appears to be the result of a misunderstanding and the ensuing damage to the public is not of such a character as to require immediate removal of the goods from the market, it is the practice of the bureau before initiating regulatory action to give notice to the trade, advising that on or after a certain date legal action under the food and drug act will be instituted if continued violations are encountered. Where the facts seem to warrant it such notice may be preceded by a public hearing at which interested parties are accorded opportunity for free discussion . . . it is the bureau's theory that more is to be accomplished by acting in an advisory capacity under such conditions as will insure legal products than by accumulating a record of successful prosecutions with attending fines turned into the Treasury of the United States."[26]

Wiley was no longer associated with the Bureau of Chemistry at the time of the Dunbar statement, having resigned in 1912 largely because of frustrations met in enforcing the Food and Drugs Act. Many food manufacturers had shown themselves more resistant to Wiley's fatherly admonishments than he had anticipated, but the chief chemist was prevented from taking sterner measures against these recalcitrants by what he considered the interference of Secretary of Agriculture Wilson. Wiley finally left the Bureau to join the editorial staff of *Good Housekeeping*, from which post he continued his campaign for a nutritious and wholesome food supply for another seventeen years. As these years passed, Wiley's bitterness at seeing full enforcement of "his" act thwarted by concessions to manufacturers grew, so that the eventual Bureau philosophy of educational correction before punishment (implicitly sanctioned by his fond father analogy of 1906) came to seem to Wiley a perversion of his life's work. But a year before his death, the old crusader charged that government officials had "read into the law . . . a meaning absolutely foreign to its purpose. The Bureau of Chemistry as constituted at the time of enforcement of the act was solely concerned in enforcing its punitive regulations. It did not consider it advisable to waste energy from its sworn duty in setting up a kindergarten or Sunday School to persuade violators of the law to desist. The law pointed out exactly what it should do, and for a short time only was this purpose of the law carried out. There is no wonder that the administration of the food law has so hopelessly broken down."[27]

Administration of the Food and Drugs Act had hardly broken down, but it had changed considerably since Wiley's departure from the Bureau. In addition to a steady shift in enforcement philosophy from punishment to education of food adulterators, there had been changes in personnel and organization. Wiley's shoes had not been easy to fill, and his permanent replacement was not selected for nine months after his resignation. Dr. Carl L. Alsberg, appointed Chief of the Bureau of Chemistry in December 1912, was a biochemist who lacked both Wiley's administrative experience and his crusader's zeal, but who had a profound appreciation of the applicability of basic scientific research to the practical problems of food and drug regulation. "Under his direction basic facts on the normal composition of foods were ascertained. Data so acquired have been of

117

the greatest value in formulating food standards which are so essential for the effective enforcement of food laws. New and better methods of analysis for detecting adulteration in foods and in drugs were developed. The operations of both Federal and state food and drug law enforcement officials were placed upon a firmer foundation through Alsberg's increased emphasis on science."[28]

It was during Alsberg's reign also that the functioning of the Bureau's regulatory force was made more efficient by reorganization. Administration of the Bureau's food inspectors had been centralized in Washington during the Wiley period, but in 1914 Alsberg divided his forces into three districts, administrative units covering the eastern, the central, and the western states respectively. Food inspection activities in each district were directed by a district chief, whose territory was further divided into various stations. This system of organization remained in effect throughout the period under consideration in this book.

Alsberg resigned as Bureau chief in July 1921 and was succeeded by Walter G. Campbell, a man quite different in background and abilities.[29] Campbell's experience had been not as a scientist (he in fact insisted on the title Acting Chief because he felt the Bureau's permanent chief should be a chemist), but as an inspection officer. Entering the Bureau in 1907, Campbell, with his exceptional abilities, became Chief of the Eastern Food and Drug District during the Alsberg administration.[e] He served as the Bureau's act-

---

[e] Campbell was destined to lead the Bureau. After receiving the highest score on the Civil Service examination given in 1907 to more than a thousand prospective food and drug inspectors, he and the thirteen other successful candidates were gathered in Washington in May to be introduced to their new duties by Wiley. At the end of the tiring first day, the Bureau Chief, relaxing in his office, called out to his secretary:

"Molly!" he said suddenly. "Do you know who's going to run this show?"

"Run it! What do you mean?"

"I don't know what his name is, but he's that tall, black-headed fellow who stood in the doorway."

ing chief until October 1923 when the eminent chemist Charles A. Browne was appointed chief. Campbell's regulatory skills were too valuable to be abandoned, though, and his resignation as acting chief was immediately followed by his promotion to the newly created position of Director of Regulatory Work in the Bureau of Chemistry. Campbell was thus to remain in charge of enforcement of the Food and Drugs Act for the next twenty-one years, first in the Bureau of Chemistry, later as director of the Food and Drug Administration. Campbell had recognized during his days as acting chief that the dual functions of the Bureau of Chemistry—those of agricultural research and of enforcement of the Food and Drugs Act—could be better performed by two separate agencies. Eventually, on July 1, 1927, his scheme of reorganization was actualized by the dissolution of the Bureau and the transfer of its chemical research activities to a new Bureau of Chemistry and Soils. The old Bureau's regulatory functions were vested in the specially created Food, Drug, and Insecticide Administration,[f] renamed simply the Food and Drug Administration (FDA) three years later. The FDA significantly remained within the Department of Agriculture until 1940, however, so that activities such as spray-residue con-

"Oh," Mrs. Read said, "That's Walter Campbell! Don't you remember, Dr. Wiley? He's the one who passed at the head of the list."

"Don't give a damn if he did," said Dr. Wiley. "I still like his looks." And the next day he appointed him chief inspector. (Related by Ruth Lamb, *American Chamber of Horrors*, New York, 1936, p. 283.)

[f] The term "insecticide" was included in the agency's original title because one of its regulatory duties was to be the enforcement of the 1910 Insecticide Act. Foods were not the only items that could be adulterated. Almost from the date of its introduction as an insecticide, Paris green had been diluted with cheaper inert ingredients by some unprincipled merchants. Several states passed ordinances to punish the practice before the federal government enacted its prohibition of the adulteration and/or misbranding of insecticides and fungicides. The penalties under this law, incidentally, were identical to the criminal penalties of the Food and Drugs Act. The Insecticide Act had previously been enforced by an Insecticide and Fungicide Board.

trol continued to be hindered by the general Department emphasis on promoting agricultural prosperity.

That the FDA, and earlier the Bureau of Chemistry, developed a serious concern for the health hazard of spray residues has been suggested in a number of places in the preceding pages, but no details of the federal government's involvement with this form of food adulteration have been given. It has been necessary to preface any examination of the specific problem of insecticide residues with a broader discussion of the evolution of federal food regulation philosophy and policy. It should by now be clear that enforcement of the Food and Drugs Act was never a matter of simply translating the letter of the law into direct action. Charting the course of regulation required careful consideration of the prevailing political and economic winds. Particularly in the case of spray-residue control, the Bureau found itself in the predicament of Gilbert's "Pirate 'prentice," torn between duty and expediency, and following one course of action while claiming to desire another. The analogy, however, is not complete, for strict adherence to duty, the Bureau soon learned, was an ideal much more easily attained in the fantasy of *opéra bouffe* than in the real world of food and drug regulation.

At least some Bureau of Chemistry personnel had been aware of the existence of spray residues since the assumption of food regulatory duties. In 1907, the first edition of Wiley's popular treatise on *Foods and Their Adulteration* had been published, and had included mention of adulteration by arsenical insecticides. The Wiley reference to residues, appended to his discussion of apples, appears to have been added almost as an afterthought. He presented residues as an unavoidable accompaniment to the necessary practice of spraying fruit, and implied they were of little consequence for health. He did advise that any apples to be eaten raw should be peeled, but neglected to explain that arsenic was the poison thus removed. The phrasing of Wiley's discussion of apples, in fact, might easily have left his readers with the impression that the most serious form of adulteration of the fruit was "the attempt which is sometimes made to deceive the purchaser . . . by placing the best . . . fruit on top."[30] These remarks in support of arsenical spraying, furthermore, were made only three pages before the denunciation (quoted before on pp. 109-110) of the bleaching of fruit with sulfurous acid and the addition of toxic chemicals to food for cosmetic purposes generally. The indication is that the Bureau chief considered spraying to be necessary for more than appearance's sake, and with so many other seemingly more urgent forms of adulteration to combat, and with funds and staff available to combat them being limited, it is not surprising that spray residues attracted no more serious attention from the Bureau of Chemistry until 1919.[g]

To be sure, the Bureau's W. D. Lynch, in collaboration with several scientists from other Agriculture Department bureaus, did initiate in 1915 a thorough study of arsenic and lead residues on produce. Their intention, if objectionable residues were found, was to modify the Department's recommended spray schedules so as to reduce residue levels. But the results of this study were not published until 1922, and it was then concluded that spray residues would not accumulate to dangerous amounts if the Department's established spraying recommendations were followed.[31]

g The Bureau was nevertheless actively involved with insecticides during the period from 1910 to 1919. A number of documents relating to prosecutions of chemical manufacturers for adulteration of insecticides in violation of the 1910 Insecticide Act can be found in the Bureau's files for this decade (the Insecticide and Fungicide Board included a Bureau of Chemistry representative). Bureau personnel also engaged in studies of the effectiveness of several varieties of arsenical sprays, and

In the meantime, the City Health Department of Boston had discovered that farmers apparently did not always observe recommended procedures, and that fruit with excessive residues did sometimes reach the market. The Bureau of Chemistry's reaction to this discovery, as indicated earlier, was to organize a conference of regulatory personnel, executives from other Department of Agriculture agencies, congressmen from western fruit-growing states, delegates from the Boston Health Department, and representatives of the fruit industry. By drawing all concerned parties together, the Bureau hoped to thoroughly air the residue problem and arrive at a cooperative program of action to reduce arsenical residues without resorting to legal actions against fruit growers. The confidential minutes of this conference on "Arsenic on Apples and Pears, Caused by Spraying," make fascinating reading, for they show the conference participants unwittingly going through what was essentially a rehearsal of a debate which was soon to become public and to be given an extended run.[h]

Several leitmotivs shaped the conference's deliberations, dominant among them being disagreement over the toxicity of spray residues.[32] Boston's Dr. Woodward disclosed that fruit seized by his Health Department had carried as much

---

of arsenical injury to plant foliage and orchard soils (see *Federal Food, Drug, and Cosmetic Law. Administrative Reports. 1907-1949*, Washington, 1951, pp. 329, 333, 396).

[h] The first seizures of sprayed produce were made by municipal authorities, and the significant deliberations over how best to ameliorate the hazard were conducted at the federal level. There was also concern for spray residues among state officials responsible for supervising the intrastate marketing of food, though in too many areas state authorities apparently neglected the problem of residue contamination (for discussions of residue regulation at the state level, see W. C. Geagley, *American Journal of Public Health*, 26, 374-376 [1936], and W. F. Cogswell and J. W. Forbes, *ibid.*, 379-381). As the states essentially merely understudied, and often without much enthusiasm, the federal role in residue control, their activities will not be discussed in any detail.

as 0.02 grain of arsenic per individual apple, and up to 0.08 grain of arsenic per single pear. The "world tolerance" for arsenic, it will be remembered, was 0.01 grain per *pound*, but this tolerance was not mentioned during the conference, nor did Woodward's figures strike all his listeners as alarming. Bureau Chief Alsberg, presiding over the meeting, observed that the maximum residues were comparable to medicinal doses of arsenic and thus presented the risk of chronic intoxication to more susceptible members of the public. Yet J. W. Summers, congressman representing the apple state of Washington and a man trained as a physician, challenged "anyone in the room" to authenticate "an acute or chronic case of arsenic poisoning resulting from the use of apples or pears." Woodward could only reply that although he knew of no such cases on record, this was probably because "the man, woman or child that has a little intestinal irritation does not suspect arsenical poisoning and, in the absence of arsenical poisoning suspicions, doctors do not suspect and, even if they do suspect, they would treat the patient and it would disappear, so that it is beyond the question to ask anyone to tell of a case of arsenical poisoning before we take action to prevent such arsenical poisoning."

Such tentative toxicology could not sound convincing to fruit-industry sympathizers. A Bureau of Plant Industry representative announced, "I do not believe it is poisonous at all; I am willing to eat any of these apples, personally." Representative Summers even suggested, "a little arsenic is beneficial," an opinion derived from the belief held by some physiologists at the time that arsenic is a normal constituent of the human body, an ingredient essential to normal physiological function.[33] This belief was based on the observation that all human bodies contain at least traces of arsenic, and indicated a physiological need regularly to ingest that substance, with food or otherwise. Alsberg appreciated that the presence of arsenic in all people is accidental, a result of the poison's ubiquity in the modern environment, and facetiously replied to Summers that "we better get that [arsenic] from Fowler's solution." No sooner had the laughter subsided, though, than the determined Dr. Summers rejoined, "Well, I think we better take that as *nature*[1] gives it to us in the apples."

The representatives of the fruit industry shared this skepticism of the harmfulness of residues, but were more concerned at the conference to develop two other themes. Fruit producers were first of all resolved to establish their own innocence of any wrongdoing. Spraying, they protested, was not simply a method they had chosen as the most effective defense against insect pests, but was a practice urged upon them for years by the federal Department of Agriculture. In the Pacific states, moreover, growers were legally "compelled to spray, because we are not allowed to ship out apples affected by coddling [sic] moth." Farmers had thus been obediently following state and federal policy in using arsenical sprays, and now suddenly a governmental agency was threatening to confiscate sprayed produce as poisonous. "The grower is in this anomalous position," Senator George Chamberlain of Oregon complained, "that if he follows your prescription he may find himself in a position where he cannot sell his fruit and, if he doesn't follow it, he will be in trouble." Fruit growers had some right to feel betrayed by their own government, and one even suspected an ill-defined conspiracy. His suspicion that "either this Bureau or the Health Department of Boston wants to get up a hue and cry that will destroy millions of dollars" was not too seriously received in 1919, but it was eventually to spread and infect the minds of thousands of angry western farmers.

At no time did the suspicion of conspiratorial activity against western agriculture have any foundation. As Woodward remarked, "If we had desired to do that ['destroy millions of dollars' belonging to the fruit industry], it would have been very easy." But it was precisely a potential loss

[1] Emphasis mine.

of money that was the second focus of the comments made by agricultural representatives at the conference. How, they asked, could growers produce fruit free of arsenic and still sell it at a profit? If arsenical sprays were abandoned, high yields of top quality fruit would become an impossibility. If arsenicals were retained, sprayed fruit would have to be cleaned before shipment, a procedure requiring considerable time and labor. In either event, the price of good fruit could be expected to soar, perhaps as high as the twenty cents an apple predicted by a farming delegate from Oregon, and the market would drop. Government representatives tried to dispel the forebodings of economic catastrophe by arguing that cleaned fruit was more appealing to the consumer than that still covered with powdery residue, and the costs of cleaning would be offset by increased sales. Further, cleaning should not even be necessary for produce sprayed according to Department of Agriculture specifications. Only where fruit was sprayed too often and too late in the season, the government contended, would objectionable residues be found. The whole regrettable Boston affair was thus the consequence of a few growers getting "a little careless."

The few fruit growers present at the conference begged to differ. Western farmers, they maintained, had not been careless, and had not oversprayed. Only a small percentage had harvested arsenical fruit, and these had all been victimized by the weather. "As the result of an extremely dry season . . . this spraying of arsenic has not weathered off," and, should similar climatic conditions recur, cleansing of harvested fruit would indeed become necessary.

Even more distressing than the possibility of having to remove residues, however, was the vision of the heavy losses that would follow "any unfortunate publicity." The public was not generally aware of the existence of arsenical residues, but if word of the Boston seizures or the Washington conference should be leaked to consumers, it was feared, they might panic and at once forswear fruit altogether.

"Unfortunate publicity" of the residue hazard could thus cause farmers to pay with their livelihoods for a crime they had not known they were committing. Alsberg agreed this injustice must be avoided, and, in the process of elaborating on the point, outlined the spray residue policy to be followed by his Bureau for the next several years: "I think what is necessary is for a great deal of publicity to be given, not in the markets or the consuming centers, but among the growers of apples in the various sections where there has not been much rain, so that everybody that ships apples will know of this situation. . . . What we would like, of course, is to make a survey of that section, out there, in cooperation with the Bureau of Markets and the state agencies, to see what the situation is and to get the situation before the people, out there, so that they may take whatever steps are possible to get their fruit in shape."

In less diffuse terms, what the Bureau of Chemistry committed itself to at the 1919 conference was a policy of educating fruit growers to the existence of a residue problem and advising them on how to hold residues to acceptable levels, while avoiding any publicizing of the problem outside of agricultural circles. It seems to have been sincerely believed that such a policy would work, that it would allow farmers time to adjust to their new responsibilities, and that this time period would be so short as to impose no risk on the unknowing public. Several Department of Agriculture officials were so optimistic as to dismiss even the educational programs as unnecessary, offering the opinion that "the lesson taught by last year's [1919] experience would stir up all the shippers to watch the fruit this year and that there would probably be no trouble."[34]

Nevertheless, to insure there would be no trouble, inspectors in the Bureau of Chemistry's Western District visited fruit growers and shippers in the Pacific states before the 1920 season, and urged them to wipe clean all fruit bearing visible spray residue.[35] These efforts were reinforced by a Department of Agriculture circular issued in early summer

1920, for publication in the agricultural press in all the major fruit producing states. It warned farmers, "DO NOT OVERSPRAY FRUITS":[36]

"Growers of apples and pears, especially, are cautioned . . . against excessive applications of spray mixtures for the second brooding of the codling moth. . . . The high market value of fruit products warrants careful attention on the part of all growers to the various orchard operations and especially to spraying. The Department feels, however, that necessary protection from pests can be secured without danger of leaving spray residues on fruits at harvesting. Careful directions for spraying have been given by the Bureau of Entomology and the Entomologists connected with the agricultural colleges in various states which, if followed, will protect fruit from insects without leaving an objectionable residue. Some injury to the fruit growing industry resulted last year from the seizure by certain health officials of fruit showing considerable spray residue. While it is generally recognized that spraying is an absolutely essential part of fruit growing, the Department feels that carelessness in making the summer applications may result in undeserved loss to growers of fruit through fears of the public caused by the few exceptional instances when over-sprayed fruits may reach the markets."

The circular was clear, frank, and, above all, conciliatory. It recognized the farmers' need to spray, and implied that only a minority sprayed carelessly. It offered simple directions on how to avoid excessive residues, and warned of the economic consequences of failing to follow these directions. Yet the circular, and related educational efforts, failed. In the late summer and early fall of 1920, Boston health authorities again made seizures of western apples and pears for bearing arsenical residues comparable to those of the previous season.[37]

The Bureau of Chemistry called another conference. The chief product of the renewed deliberations was a more explicit formulation by the Bureau of the policy it would observe with regard to residue violations: "It was then determined that seizures of . . . excessively sprayed fruit would be made only in extreme cases. In place of seizure it was determined that where objectionable shipments were discovered the shipper or consignee would be notified and given an opportunity to cleanse the fruit by wiping or to turn it over to some canner or preserving establishment which would agree to peel the fruit, thus removing the excessive arsenic. Action by way of seizure would be taken only in those cases in which the growers or shippers failed to take advantage of this privilege."[38] Bureau officials agreed that although seizure and destruction of contaminated fruit was sanctioned by law, such action was to be shunned as "an unwarranted waste of valuable food."[39] It still seemed reasonable that the fruit industry could be gently persuaded to observe the new residue restrictions, and Bureau personnel were again dispatched to the Pacific states to advise growers on residue removal, and in some areas even to provide gloves for wiping the fruit clean as it was picked.

The results continued discouraging. In the summer of 1921, Bureau Acting Chief Campbell complained of the "little impression which has been made . . . in the northwest by the efforts of the different Bureaus [members of the Bureaus of Plant Industry and of Entomology were also involved in this program] of the Department to impress upon growers and shippers the necessity for the employment of care in the distribution of fruit bearing excessive arsenic spray."[40] Only four months before, the Boston Health Department, for the third consecutive season, had detained, and eventually destroyed, a carload of apples from Washington.[41] At about the same time, the Los Angeles Board of Health found it necessary to destroy locally grown arsenical celery, and to take similar action later in the year against sprayed apples.[42] Several illnesses were attributed to both the celery and the apples, but through the efforts of local and state commerce agencies, these developments were kept out of the newspapers.[43] The Bureau of Chemistry's Western District personnel followed the events in Los Angeles closely, but could take no action since the produce involved had not been moved in interstate commerce. Nevertheless, it was becoming clear that Bureau policy was not producing the anticipated results and that a more rigid posture would have to be assumed. Campbell informed a Washington state entomologist in the fall of 1921 that: "The Bureau believes that the educational work which has been done during the past seasons has been sufficient to acquaint all shippers and growers with their responsibility. We do not believe that further leniency by way of refraining from seizure action, with the attendant undesirable publicity, will be justified should further shipments bearing excessive arsenical spray residue be made."[44]

Its mood was threatening, but while local health agencies continued to make isolated seizures of sprayed produce over the next several seasons, the Bureau clung to its low-key policy of persuasion until 1925. To be fair, the Bureau's forebearance was in effect dictated by the economic and political realities of the early 1920s. In the spring of 1920, American farmers were enjoying their greatest prosperity in years. The food requirements of a Europe devastated by war had stimulated agricultural production in this country to an all-time high, and as long as this demand continued, crop and livestock prices rose. But when the European market began its inevitable decline, in the summer of 1920, the American farmer became an over-producer, and the immediate collapse in agricultural prices plunged him into depression nearly a decade before the stock market slump lowered the rest of the country to his level. During that decade, more than two hundred thousand farm homes were abandoned,[45] in spite of the organized efforts of congressmen from rural states to stem the agricultural ebb. The Farm Bloc, organized in 1921, was a coalition of representatives and senators from the South and Midwest that actively campaigned for legislation helpful to farmers, and actively opposed legislation considered injurious. The bloc was heartily encouraged in its organization and activity by Henry C. Wallace, Secretary of Agriculture from 1921 until his death in 1924. As Secretary of Agriculture, furthermore, Wallace had to approve of any regulatory actions requested by the Bureau of Chemistry. With such deterrents facing it, the Bureau's laxness during the early years of residue regulation is understandable; indeed, it calls to mind Samuel Johnson's remark about a dog walking on his hind legs: "It is not done well; but you are surprised to find it done at all."

Throughout these early years of regulation, however, the Bureau's patience and sympathy with the farmer were being steadily eroded by continued spraying abuses, until they wore dangerously thin. In the spring of 1925, Florida was invaded by the celery leaf-tier, a yellow-green caterpillar that thrives on celery foliage and that in sufficient numbers can destroy a whole region's celery crop. Knowing this, Florida celery growers drenched their crops with too concentrated mixtures of lead arsenate, applied repeatedly almost up to the day of harvest.[46] When Bureau of Chemistry inspectors became aware of the unrestrained spraying, they notified growers of the legal actions to expect if they did not modify their spraying practices and clean their celery before shipment. The growers at once promised obedience but in April inspectors found marketed samples of celery containing as much as 1/7 grain of arsenic per stalk! Bureau officials now firmly warned celery shippers, on pain of prosecution under the Food and Drugs Act, not to market the vegetable until its arsenic was removed. It is clear from Bureau records that seizure actions were imminent, but the shippers finally satisfied the government's demands and prosecution was avoided. The Bureau had nevertheless shown itself ready to strike back if pushed too far.

In August of the same year, there appeared in Philadelphia newspapers alarming stories of fatal poisonings from New Jersey fruits.[47] Fearing that spray residues had finally claimed some victims, W.R.M. Wharton, the Bureau's Chief of the Eastern District, conducted an investigation of the alleged poisonings. It was learned that epidemic illness characterized by symptoms similar to those of arsenic poisoning was indeed occurring in southern New Jersey, but that physicians doubted that their patients' problems were caused by arsenical fruit. Among those stricken, for instance, were children too young to eat fruit, and adults who had not eaten fruit recently. Further, intestinal influenza had been common in the state during the summer, and the symptoms that journalists were attributing to arsenic actually coincided more closely with influenza.

In the process of clearing sprayed fruit of the blame for this epidemic, however, Wharton's inspectors found that many New Jersey apples carried "amazingly high" residues of arsenic, as much as a medicinal dose on a single apple in some cases.[48] New Jersey orchards had been assaulted by Japanese beetles that season, and some growers had responded by spraying heavily, as many as twelve times, and close to the date of harvest. As a result, the possibility of an actual epidemic of arsenic poisoning seemed no longer so remote, and the Bureau immediately notified the New Jersey State Department of Health of the situation.[49] In the meantime, Bureau inspectors began making the rounds of New Jersey orchardists and warning them that their produce was contaminated and, if shipped in that condition to another state, could be confiscated and destroyed. Another conference was even convened in Washington, at which government representatives tried to convey to New Jersey growers some feeling for the gravity of the residue problem. "It was pointed out at this meeting," one of the participants recorded, "that the Bureau of Chemistry was not anxious to make seizures with the consequent possible damage to innocent growers, and that the Bureau of Chemistry would lend as much assistance as possible to correct the situation by persuasive methods."[50]

Persuasion again failed. In November, inspectors from

the Philadelphia Station of the Eastern District, maintaining "a continuous surveillance over all ferries crossing from New Jersey into Philadelphia," arrested several New Jersey growers for bringing contaminated apples across the state line. More than 150 bushels of fruit were seized by the Bureau, and when the owners were unable to clean it satisfactorily, the fruit was destroyed.[51]

This was the first actual seizure of sprayed produce for violation of the Food and Drugs Act, and the placatory nature of the Bureau's action is again noteworthy. State health officials in Pennsylvania had arrested and prosecuted dealers who sold contaminated New Jersey fruit in Philadelphia, and under the provisions of the 1906 Act, the shippers of the objectionable apples were subject to a federal penalty of a two-hundred-dollar fine, in addition to the loss of their produce. In keeping with its past policy, the Bureau allowed New Jersey growers more than three weeks to attempt to clean their fruit before it was destroyed, and brought no criminal action against the growers themselves. As a Bureau official explained to the growers: "The Bureau is interested in preventing the merchandising of this fruit which contains excessive arsenic. It is not interested in prosecuting citizens of New Jersey and I may say to you that if the problem can be solved without the necessity of bringing these prosecutions and if prosecutions are not needed to prevent the recurrence of these shipments in this or other years, the Department will probably not prosecute."[52]

The Bureau seemed incurably optimistic. Just a few weeks before the New Jersey seizures, however, another incident involving sprayed apples had set in motion events that would force the Bureau of Chemistry to reassess its policy and develop a more concrete and more realistic program for handling the spray-residue problem.

*London office advises trade rumors British
Government considering placing general import
embargo on American apples unfounded* STOP
*Rumors probably originated action British
Government stopping certain shipment Oregon
apples too freely sprayed with arsenic resulting
case arsenic poisoning* STOP . . . *General
attitude British authorities is confine action
to particular shipper and brand through which
poisoning developed rather than embargo all
shipments* STOP *Cooper Bureau Agricultural
Economics advises guarded dissemination only
no general publicity.*

> —Henry Chalmers, Chief, Division of
> Foreign Tariffs, Department of Commerce

Regulatory Perplexities   5

THE RADIOGRAM of November 16 broke the news that was
to cause so much anxiety for American apple growers during the last weeks of 1925 and the first months of 1926.
Great Britain, one of the major markets for America's sur-
plus fruit, had again been offended by American apples.
Late in the nineteenth century, it will be recalled, the British agricultural press had charged that apples imported
from the United States regularly arrived with a powdery
coating of arsenic and were deadly. But as no cases of serious poisoning from foreign fruit were authenticated, the
furor soon subsided.

In October 1925 a family of four in the London borough
of Hampstead fell ill with what was diagnosed as arsenic

poisoning.[1] All foodstuffs in their home were promptly analyzed, but the only items found to contain arsenic were several apples (carrying 1/15 grain of arsenic apiece!). The family had eaten other apples from the same bag before becoming ill. Traced to the local stand where they had been purchased, the apples were found to be part of a shipment from the Western United States. Immediately inspections and analyses of American-grown fruit were instituted throughout England, with apples frequently being discovered to hold concentrations of arsenic exceeding 1/100 grain per pound, and even as high as 1/7 grain per pound. A number of British fruit retailers were arrested and fined the equivalent of twenty-five to one hundred dollars for selling the contaminated apples.

British newspapers, meanwhile, were running such headlines as "Arsenic Apples," and "Prussic Plums," and one, referring to a recent "Eat More Fruit" campaign (a publicity project of English produce wholesalers, launched before the Hampstead poisonings), commented "EAT MORE FRUIT: You will be Dead Certain if You Do."[2] By early 1926 the London office of the Bureau of Agricultural Economics was able to send the Department of Agriculture more than nine hundred articles, editorials, and cartoons from British newspapers, all critical of American fruit and fruit growers.[3]

As a result of such widespread concern, the British Government was indeed giving serious consideration to an embargo of all American fruit. In January 1926 London requested Sir George Buchanan (who had served as Secretary to the Royal Commission on Arsenical Poisoning more than two decades before) to confer with U.S. Department of Agriculture officials and attempt to find some solution short of an outright embargo. The Department assured Buchanan that every endeavor would be made to prevent the exportation of fruit bearing arsenic residues above the British tolerance, and Britain agreed to resort to a general embargo only if conditions did not soon improve.

Conditions did improve, but not as rapidly as the British

might have liked, and the American apple industry was kept nervous for several months by the threat of embargo. Its uneasiness was compounded by a decline in English apple sales. The English public had been made afraid of American apples, and wholesalers were having to reduce the price of the fruit by as much as fifty cents a box in order to sell it.[4] The American industry's chief competitors, Canadian and Australian apple growers, were exploiting the situation by advertising analyses showing that their fruit contained no arsenic. These tactics suggested to one American agricultural editor that the whole affair was a conspiratorial "stunt" engineered by the countries of the British Commonwealth "to discourage the purchase of American apples."[5]

Most supporters of the American apple industry took a more realistic approach to their troubles. The Secretary of the International Apple Shippers' Association (an American organization) later reported: "During all of this time our own representative in England . . . worked night and day for weeks and months to hold the situation from getting beyond even the hope of solution. The wholesale trade even paid retailers' fines and costs, withdrew contaminated lots from retail sale, conducted independent analyses and investigations, and . . . did everything within human power to prevent outright disaster. The rank and file of persons in this country engaged in the apple and pear industry have no idea of the tremendous load that had to be carried and the problems which had to be met. . . . The fact that the season ended without an outright embargo is a tribute to the efficiency of the work accomplished."[6]

The apple shippers' sacrifices in behalf of their industry were all the more noble for having been unsung. Their work was carried out in the strictest secrecy, for publicity, it was realized, might make the American public as jittery as the English and destroy domestic apple sales. Thus while the Apple Shippers' Association conducted a "splendid educational campaign" throughout the country to alert growers to British sanctions against excessively sprayed fruit, "all of this work has been conducted on the personal contact basis and with the strongest possible emphasis against newspaper, magazine, radio or any similar publicity. In fact the matter has been kept out of even the small sheets at country points because it is only a step from the four corners to the columns of the metropolitan dailies."[7]

The campaign of secrecy succeeded. One metropolitan daily, the *New York Times*, published a short announcement on November 26, 1925, that the British "Find Arsenic in American Apples," and the article was read by an official of Houghton College, who wrote the Department of Agriculture to inquire if a recent outbreak of intestinal disturbance among his students might be due to the cafeteria's apple pie.[8] Otherwise, the conflict between American growers and the British government escaped the notice of the American public, and no nationwide fears of arsenic poisoning were yet aroused.

As a final step to reestablish their business, the Apple Shippers' Association requested a conference with Department of Agriculture officials to work out a cooperative program for lowering arsenic residues on fruit exported to Britain. The conference was held on April 12, 1926; "complete harmony prevailed"; and by the end of the day shippers and officials had agreed to a nine-point remedial plan which included as its most significant feature provision for "Orchard inspection in many leading sections thirty days before harvest by state officials and local committees and a re-inspection at harvest in conjunction with federal officials. In these sections, orchards that are found not up to legal requirements will be embargoed and the fruit not permitted to be packed or shipped until put in condition. In addition many sections have provided for packing house and platform delivery inspection and inspectors have been clothed with the necessary powers by their respective states."[9] State and federal officials thus collaborated with the agricultural industry to insure that no fruit potentially harmful to consumers would be shipped to England. Occasionally items bearing excessive arsenic slipped through the inspection program and aroused complaints from England, but these occurrences were too few and far between for the threat of a British embargo to seem very ominous after 1926.

The irony that fruit sent to British subjects was receiving closer scrutiny from American officials than that intended for domestic consumption was not lost on the Bureau of Chemistry. Export fruit was required to pass the world tolerance for arsenic, while fruit kept in the United States was allowed to carry considerably more. As late as December 1926 a Bureau representative admitted that fruit bearing as much as four times the British tolerance for arsenic had been permitted on the American market.[10] Two months later, Walter Campbell himself tried to shame the fruit industry into cooperation with his Bureau by announcing: "Heretofore, the Secretary of Agriculture, in recognition of the needs of the industry and appreciating that a drastic [domestic] enforcement of the 0.01 grain tolerance would result in disaster to the industry, has assumed the risk of stultifying himself before the consuming public by observing an informal tolerance considerably more liberal than is justified by the physiological facts."[11]

Just how liberal this "informal tolerance" was is not entirely clear, for Bureau officials carefully avoided publicizing it, even in intra-Bureau correspondence. Chief Alsberg explained this secrecy in a 1921 letter to Boston Health Commissioner Woodward:

"We do not expect to announce any limit or tolerance for arsenic, believing that this would be inadvisable for a number of reasons. In the first place, it would have a tendency to give growers and shippers the impression that the matter of arsenic contamination is trivial, causing them to abate the efforts which they are now making to insure the production of fruit which is free from objection. If a limit were announced it would necessarily have to be made extremely stringent and considerably lower than the limit on which as a practical proposition we would feel justified in attempting regulatory action since, as you realize, for successful prosecution it will be necessary for us to demonstrate that the fruit bearing the amount of arsenic on which action is taken is actually injurious to health. If an extremely stringent limit were announced it would be interpreted literally by a great many local and state officials, and cases might be instituted which when brought to prosecution could not be sustained, thus leading in the long run to more objectionable conditions than now exist. We have for our own purpose of administration adopted a working tolerance below which we do not propose to take action. We believe that fruit containing more than this working tolerance of arsenic should be seized and feel that we can maintain cases upon it in the Courts. While it is not our purpose to make this tolerance known, I have no objection to advising you of it confidentially for your own personal guidance and information, should you desire. I do not feel that it would be advisable to communicate this working tolerance generally to the rank and file of food control authorities for the reason that it would not be long before it became known to the trade."[12]

C. W. Crawford, a chemist in the Bureau, read the letter to Woodward before it was mailed, and in a memo to Alsberg remarked, "It strikes me as peculiar that we tell him we don't object to his knowing our tolerance, and then fail to give it."[13]

Woodward was not alone in his effort to pry the tolerance loose, yet in its replies to not only other health officials, but also to concerned agriculturalists, entomologists, insecticide manufacturers, and congressmen, the Bureau revealed no more than that fruits and vegetables often carried residues of arsenic comparable to a medicinal dose of the substance, and that such residues were dangerous for general consumption. The implication was that produce bearing near-medicinal doses of arsenic would be kept off the market by the Bureau, but no precise figure for objectionable residues was ever released beyond the Department of Agriculture. The standard medical dose, 1/30 grain, was considerably greater than the 1/100 grain allowed by the British tolerance per pound of food. It was presumably large enough to impress a jury, should any seizures of produce be contested, and some figure of the order of 0.04 grain per pound would seem to have been the Bureau's initial secret working tolerance.[14]

The successful keeping of the secret did little to alleviate the residue problem. Bureau officials seem at first to have expected the marketing of oversprayed produce to be a short-lived annoyance (perhaps another reason for their failure immediately to establish a full and open tolerance policy), one that would disappear once farmers were alerted to the necessity of observing spray schedules and to the terrible economic consequences of a public discovery that its food contained arsenic. Hence the educational campaign begun so enthusiastically in 1920 had been cut back significantly by the 1924 season "since it was believed that earlier efforts along this line had produced the desired results." But "the situation in connection with the 1925-1926 apple crop was . . . especially disappointing and indicates the need of constant educational work among growers as to the relation of the Pure Food and Drugs Act to spraying operations."[15]

The unexpected situation also indicated that the policy of an informal but secret tolerance was ineffective, and that an exact tolerance publicized to the agricultural industry and firmly enforced by the Bureau might serve as a more efficient deterrent to spraying abuses. The objections of British health authorities during the 1925-1926 season similarly impressed the Bureau with the need for a reexamination of its domestic residue regulation.[16] Finally, produce growers and shippers, the people most directly affected by the Bureau's tolerance policy, were also urging the adoption of a definite and advertised (to the industry) tolerance.

An Oregon correspondent summarized the vexations of farmers the country over: "The existence of this secret tolerance, Mr. Secretary [of Agriculture], is so un-American and undemocratic that I marvel that you would for a minute tolerate it. . . . Here is a law, drastic and penal in its nature, and yet the citizen may not know when he is violating the same. It would be just as rational we say to pass a speed law with a secret limit which was to be known only to the traffic officers."[17]

The combination of these factors with the knowledge that its secret tolerance had been too lenient had pushed the Bureau by mid-1926 to consider a stricter, and publicized, domestic arsenic tolerance. In late August, Bureau Chief Browne sent form letters to a number of America's leading toxicologists: "In its enforcement of the Federal food and drugs act the Bureau of Chemistry has frequently to consider the limits to which arsenic, lead, copper, zinc and other metallic impurities may be tolerated in foods. The presence of arsenic and lead upon fruits which have been sprayed with lead arsenate is being considered by the Bureau in its examination of products which are sold to the public. Any information which you can give with reference to the limits which should be set for the presence of arsenic, lead, copper and zinc in foods will be greatly appreciated. Would you consider the presence of $1/100$ grain of As_2O_3 on a single individual apple, pear, or other fruit to be a menace to the public health, bearing in mind the fact that many persons eat several of such fruit a day without removing the skins?"[18]

A limit of 0.01 grain of arsenic per individual fruit would actually be somewhat less stringent than the British tolerance, the degree of laxity depending on the number of fruits per pound (which in some cases was three or four). Most of the toxicologists who replied suggested that 0.01 grain per pound, rather than per fruit, would be a wiser tolerance, an opinion in which even the Surgeon General of the United States, H. S. Cumming, concurred.[19] The weight of this testimony was still not sufficient to press the Bureau to a decision, and a conclusion to the issue was not forced until the intrusion of yet another regulatory difficulty.

The Food and Drugs Act provided the owner of any food product seized by the Bureau the right to contest the seizure and force the government to defend its action in a court of law. As has been seen, one of the major considerations in deciding upon an administrative tolerance for spray residues was its legal defensibility. Even after adopting its liberal tolerance for arsenic, the Bureau worried that seizures of sprayed produce might be challenged and the government be unable to convince a jury that such seemingly small residues could be dangerous.

Fortunately, fruit producers were no more eager for a trial, not because they considered Bureau policy just or were not annoyed by it, but rather because they feared the adverse publicity which might be generated by a court trial. For nearly a year after the initial seizure of sprayed fruit, fruit growers had refrained from legal protest of government confiscation, and ofttimes destruction, of their property. Then the Bureau's inspection force collided with Llewellyn Banks. Banks was the owner of the Suncrest Orchards of Medford, Oregon, a man who did a considerable volume of business in apples and pears, and one who brooked no unwarranted interference with his operations.[20] Banks and the Bureau had met before, in a disagreement over the wholesomeness of some oranges he had shipped from California, and he had closed that encounter by threatening to shoot the next pure food inspector who set foot on his property.[a] It was thus with some uneasiness that inspectors in the Western District began to object, during

[a] Bureau personnel came through their confrontation with Banks unscathed, but such threats from his lips were not necessarily idle. A few years later Banks was sentenced to life imprisonment for the murder of a policeman who attempted to arrest him for stealing ballots after a local election went against a Banks candidate.

the summer of 1926, that Suncrest Orchard fruit was not being sufficiently cleaned of its residue. Banks at once detected in these early warnings collusion between the federal government and rival fruit shippers to ruin his business, and actually managed to obtain a court order temporarily restraining Bureau inspectors from examining or reporting shipments of fruit from his packing plant. The Banks ploy seemed a great success at first, but was ultimately to backfire, for while it forced the Bureau to waste time and money trying to track down Suncrest shipments after they left Oregon, it meant that seizures of these shipments might be made in legal jurisdictions far outside the state, and any court trials resulting from seizure would be conducted away from Banks's home ground. The local favoritism shown by food and drug trial juries has already been observed. Certainly Banks, a popular figure in his agricultural home state, was offering the government an advantage by requiring that his fruit be arrested elsewhere.

Nevertheless, when the first seizure was made, Banks contested it, and in November 1926 the Bureau and Banks went to Chicago for the first legal test of the validity of the administrative tolerance for arsenic. The five carloads of Suncrest apples and pears on trial were unquestionably contaminated, residues as high as 1/9 grain of arsenic on a single piece of fruit having been found. But the government had to prove that such residues were potentially harmful; taking no chances, the Bureau called up several of the nation's most renowned medical scientists for reinforcement. Men such as University of Chicago physiologist Anton Carlson, University of Wisconsin pharmacologist A. S. Loevenhart, and Boston toxicologist W. F. Boos agreed before the jury that the Suncrest residues "not only might be harmful to health, but . . . positively would be harmful to health."[21] The verdict was returned in favor of the government.

He lost the case and his fruit had to be destroyed, but the beaten Banks had set an example of boldness to stiffen other orchardists' resolve not to be abused by the government.[b] Henceforward, seizures of sprayed produce were to be regularly contested, and the Bureau's force of poisoning experts would stay busy marching from one federal courtroom to the next to present their evidence against lead arsenate. Since it is almost correct to say that to see one of these residue trials was to see them all, excerpts from the transcript of only one trial will sufficiently illustrate the tactics of government and growers:

"On or about October 23, 1926, the United States attorney for the District of Kansas, acting upon a report by the Secretary of Agriculture, filed in the District Court of the United States for said district a libel praying seizure and condemnation of 630 baskets of apples at Hutchinson, Kansas, alleging that the article had been shipped by F. L. Martin from Clifton, Colorado, on or about October 15, 1926, and transported from the State of Colorado into the State of Kansas, and charging adulteration in violation of the Food and Drugs Act. . . .

"It was alleged in the libel . . . that the article was adulterated in that it contained an added poisonous ingredient, to wit, a compound of arsenic and lead, which might have rendered it injurious to health.

"On March 21, 1927, F. L. Martin, Hutchinson, Kansas, having appeared as claimant for the property and having filed an answer to the libel denying the adulteration of the product, the case came on for trial before the court [at Wichita] and a jury."[22]

The Martin apples had been found to carry as much as 1/9 grain of arsenic apiece, a residue far above the world tolerance. Yet a more vivid picture of the dangers of arsenic and lead poisoning would have to be painted to attract jurors to the government side.[23] The first stroke was drawn

---

[b] Shortly after his return from Chicago, Banks was given a pocket piece by his fellow Medford fruit growers. It was inscribed: "A reminder of your fight against bureaucracy." Ruth Lamb, *American Chamber of Horrors*, p. 220.

by pharmacologist Loevenhart. After outlining his years of research on the toxic effects of both warfare chemicals and drugs containing arsenic, Dr. Loevenhart was asked:

"Q. Taking into consideration the testimony of the witnesses who have testified here before you as to the specific amount, the aliquot parts of these various poisons in these apples, would the general consumption, if permitted by the consuming public, would it render the consumption of the food in question injurious to health generally by reason of these poisons?

"A. I should answer your question that in the quantities in which these poisons are present on these apples, that these apples taken by persons day after day, in various forms, consumed generally as food, in the ordinary course, in the ordinary way in which apples and pears are consumed by the public, that that would constitute a very serious menace indeed to the health of the public.

"Q. That answers my question, thank you."

To drive deeper into jurors' minds Loevenhart's assertion of the danger of such small quantities of lead arsenate, the government called Dr. Charles Dennie (a member of the University of Kansas Medical School faculty and a specialist in the treatment of arsenical dermatitis) to testify that "the chronic ingestion of very small quantities of lead is fraught with great danger." From personal experience, he recalled "a number of cases in which puzzling exfoliative dermatitis has been traced to the ingestion of small quantities of lead . . . [and] also referred to the chronic lead poisoning produced in the wife of a University of Kansas professor by the use of a face powder."

Dr. Dennie was followed to the stand by Dr. J. C. Munch, Bureau of Chemistry pharmacologist, who reported on controlled laboratory experiments in which the peelings from some of Martin's apples had been fed to rabbits. Subsequent autopsies had shown significant quantities of arsenic and lead to be present in both adult animals and in young ones nourished only by their mothers' milk. In addition, the growth of the young animals had been retarded, and the internal organs of the adults irritated by the poisons. Several rabbit "stomachs and kidneys were placed in formaldehyde and offered as evidence. . . . They made a very strong impression on the jury."

The impression of residue poisoning had to be strengthened further, and Anton Carlson was the man to do it. Asked about his experience "with the solubility of arsenate of lead in the human system or in the human gastric juice," Carlson recounted his research with Fred Vlcek.

"Q. What did that test show, Doctor, or what did these tests show?

"A. The tests showed that all the lead salts, dry, were soluble in human gastric juice, but the solubility differs. . . .

"Q. Was lead arsensate soluble, according to your test, Doctor?

"A. Lead arsenate was soluble in human gastric juice according to our test. . . .

"Q. What is the effect of food in the stomach on the solubility of arsenate of lead by the human system?

"A. That depends on the kind and the quantity of food. Food like an apple would not much reduce the solubility, but food like milk or meat which would diminish the acidity of the food would decrease the solubility in the gastric juice itself. Now what happens after the food passes from the stomach into the intestines, it is more difficult to say because there is evidence that lead goes into the system, into the blood and tissues, without us being able to explain it on any basis of solubility. . . .

"Q. In your opinion, Doctor, would the organic acids in apples increase or decrease the danger of lead poisoning or arsenic poisoning from eating apples containing the amount that has been testified of lead arsenate?

"A. In my opinion, any acid, including the weak organic acids in apples or pears, would increase the solubility, being that there is virtually nothing in apples that would decrease the acidity of the gastric juice. . . .

"Q. Is it or is it not a fact, Doctor, that lead in any form . . . when taken by the mouth or otherwise administered, has proven to be a cumulative poison?

"A. That is correct.

"Q. What is a cumulative poison, Doctor?

"A. A cumulative poison is any substance which stays— either stays in the body in increasing amounts as you receive small amounts from day to day, until a quantity is reached which produces demonstrable evil effects on the processes, or a poison, I should say, which gradually weakens an organ or a process without actually accumulating in amount until an effect is produced.

"Q. Are or are not all lead compounds that have been studied poisonous?

"A. Every lead compound or metallic lead, as far as human experience or actual experiments have been made, have proved poisonous, but the degree of resistance to such poison varies greatly, and in the case of a healthy individual a great deal of internal damage must be done before the individual actually becomes sick.

"Q. Is there any reason to believe that arsenic [sic] of lead would not be just as poisonous, since it generally contains arsenic, than any other compound containing lead?

"A. No, there is nothing known.

"Q. What are the facts?

"A. There is nothing known that would make us less suspicious or less careful with regard to this particular lead compound than any other compound of lead. In addition, there is the possibility of a synergistic action, that is to say, the two poisons acting together. . . . In other words, arsenic may sensitize—I don't say that it does, to lead, and lead to arsenic.

"Q. Is there or is there not any doubt whether arsenate of lead is a cumulative poison?

"A. There is no doubt in my mind—there is no basis for reasonable doubt. There is no compound of lead known which, when getting into the system in sufficient amount, over a sufficient length of time, has not proved poisonous.

"Q. How long would it take, Doctor, to prove by experiments that arsenate of lead is a cumulative poison?

"A. It would not take long to prove that it may be an acute poison, because there you can use big doses and get quick responses. But an accumulative poison, in very small doses, would require years, and varying numbers of years probably, depending upon the individual. It would take a shorter time in the case of children and women to appear, according to some authorities at least, as they are more susceptible to lead poisoning than adult men. There are cases on record of individuals working from five to ten to fifteen years in lead industries and the lead poisoning showing up only after the end of that time. . . .

"Q. Doctor, assuming that apples or pears contain added concentrate of lead in amounts varying between one/one-hundredth and two/one-hundredths of a grain per apple or pear, and that apples and pears were eaten as a food by the strong and the weak, the old and the young, the well and the sick, in your opinion would or would not such an apple or pear be injurious to health if eaten by any of the humans above mentioned?

"A. Without question.

"Q. Over a period of time, I meant to add to the question, Doctor.

"A. Over a period of time, such a consumption of such apples would be injurious to health and if continued long enough might produce demonstrable lead poisoning."

There might remain one question for the juror even after Carlson's harrowing testimony: If lead arsenate were so common and so serious a poison, why were not its victims more conspicuous? Chicago physician Frederick Tonney was questioned by government counsel:

"Q. Doctor, would it be possible for people to be poisoned by arsenate of lead on apples and pears and the doctor not know from where the poison was obtained?

"A. I think it is not only possible, but it would be very

probable that the doctor would not realize the source of such a poison.

"Q. Doctor, as an administrative proposition, in the Health Department of the City of Chicago, do you consider these apples harmful, according to the testimony of the analyst here?

"A. I should say in my opinion apples containing the amounts of arsenic specified would be excluded from the Chicago Market by the Chicago Health Department.

"Q. For what reason?—as being injurious to health?

"A. As being injurious to health."

The same line of questioning brought even more disturbing answers from toxicologist William Boos:

"Q. Doctor, would it be difficult for a doctor, seeing a patient who had been chemically poisoned with arsenate of lead by eating apples or pears, to find out where he was getting this poison?

"A. It would be very difficult. May I give an example?

"Q. Gladly.

"A. I was called in to see two persons, who had a severe gastro-enteric ailment, both stomach and bowels, and the doctor in charge of the case . . . was at a loss to know why these two patients, who were children, respectively eight and ten years old, were so ill. He called me in in the capacity of a physician and not a toxicologist. I inquired into the causes which could possibly be held responsible for their illness and I found that these children had eaten a considerable quantity of currants which had been sprayed with arsenate of lead.

"Q. What, Doctor?

"A. They had eaten a considerable quantity of currants, small fruit, which had been sprayed with arsenate of lead and their illness was acute arsenical poisoning. That was checked up by the analysis of gastric contents and by analysis of urine, and stools. They were made acutely ill by toxic poison. The doctor had no thought of what they had eaten and this fact that it was poison would not have occurred to him as he told me. I think in the same way there are thousands of cases where people are made ill by eating these things and it is not recognized by the general practitioner because it is a wholly specialized thing to recognize the symptoms of these diseases.

"Q. Doctor, could a doctor successfully treat a chronic arsenical or lead poisoning [without knowing where the poison] had come from?

"A. It would be absolutely impossible for a physician to treat a case of chronic arsenic or chronic lead poisoning, even if he recognized them as such, if he did not know where the arsenic or lead came from, because it would not be possible for him to remove the cause and the patient might keep on taking it into his system, and he would be absolutely at a loss to safeguard him in that way. He must know the sources in order to be able to prevent further taking into the system of this poison."

F. L. Martin, the claimant, was himself a judge, and appreciated the fact that, expert and assured as all the government's witnesses were, their contention of the dangerousness of a thin residue of lead arsenate still strained a juror's belief. His counsel demonstrated with the first witness that he knew their hopes rode on reducing toxicology to an absurdity. Arguments used by entomologists for years to rationalize the safety of spraying were thus marshalled forth to confound Loevenhart, but the doctor was exasperatingly resistant:

"Q. Now, what is a lethal dose of this arsenate of lead? . . .

"A. I would say that the amount which might prove fatal in a single dose to a human being might vary anywhere from two to ten grains. . . .

"Q. Now, Doctor, assuming that a person has been engaged in spraying this fruit for a number of years, using a lead arsenate spray in putting it upon the trees with a high pressure machine; that he sprayed it all over, so that his clothes were absolutely saturated with it, and his legs covered with the lead arsenate spray . . . so that he breathed it, and took it in in various ways; that for a number of years he had been engaged in spraying this fruit, and then later, after the spraying season was over, had been engaged in the packing houses in packing and inspecting, and had constantly eaten this fruit . . . had been engaged in handling this fruit before it was wiped, and had every day eaten a number of apples and pears, would you say that a basophilic blood test should show lead in his system?

"A. I should say that if the lead entered his system, no matter what he was engaged in, if the lead entered the system in any appreciable amounts, he would show changes in the body, as everybody else is known to show it, when taking lead in sufficient quantities to produce chronic lead poisoning.

"Q. Well, in the case of a person eating this fruit in quantities for a long time that test would be applied, would it not?

"A. Yes, sir.—I don't know whether it would or not. If the doctor thought that the patient was suffering from lead poisoning he would certainly apply the test in order to verify and in order to substantiate his diagnosis that the person was suffering from chronic lead poisoning; otherwise, it would not be made, I don't think.

"Q. The test would be indicative of the amount of lead that was in his system, would it not?

"A. It would be indicative that he had lead. It would not be indicative of the amount. You cannot say, from the amount of these basophilic changes how much lead he had received. . . .

"Q. One of these apples contains merely 4/100ths of a grain.

"A. No, the testimony shows . . . that the average of arsenate of lead is approximately 1/5th of a grain. . . .

"Q. To the apple?

"A. To the pound. No, to the apple. . . .

"Q. How much arsenic does that represent?

"A. It represents approximately as I figured it out, per apple approximately 1/60th of a grain of arsenic [sic, approximately 1/20th grain].

"Q. Now, as a matter of fact, 1/60th of a grain of arsenic is very frequently given as a tonic dose to anemic and sick people?

"A. A dose—if I may just have my pencil and paper—a dose of that kind is frequently given under the doctor's direction and under the doctor's supervision to a patient. However, of course, the doctor is always watching for signs of poisoning when that dose is administered and is prepared to withdraw the dose if the patient shows any of the symptoms of chronic arsenic poisoning.

"Q. Well, there is not any probability of a person taking 1/60th of a grain of arsenic, that that person would show any symptoms of chronic poisoning.

"A. Not only is there a probability, but it occurs all the time in the medicinal use of arsenic, that the doctor finds that the patient [is] showing puffiness under the eyes, showing other evidences of chronic arsenical intoxication and orders the drug discontinued as the result.

"Q. Does he ever get that condition from 1/60th of a grain?

"A. 1/60th of a grain is a little less than is customarily used. . . . Now, it is also known that 1/60th of a grain taken day after day has caused symptoms of chronic arsenical poisoning.

"Q. Well, is it not a fact, doctor, that on the contrary, a person taking arsenic, if the dose is increased, starting with 1/60th of a grain, if it is gradually increased, the person builds up a tolerance?

"A. No, sir, I do not believe that.

"Q. If the dose is increased he builds up no tolerance?

"A. The tolerance itself is increased sometimes by as much as ten or twenty times the amount, but as soon as the patient shows any symptoms of chronic arsenical poisoning, for which the doctor is watching, the drug is entirely withdrawn. In other words, he knows the patient is getting arsenical poisoning and stops it. It is always given for given periods, never indefinite periods of time."

The line of attack on toxicology had run out with Loevenhart, as it would with the other medical experts. But Martin's defense was not exhausted; one ploy was left, revealed in Judge John Pollock's homely charge to the jury:

"Gentlemen of the Jury: This case has now proceeded to that point at which it becomes the duty of the court to charge you as to the law that will govern you upon your deliberations upon a verdict in this case. . . . Now, the object and purpose of Congress in the enactment of this pure food and drugs act was the preservation of the health of people, and a very right kind of an act. The defendant in this case admits that he raised these apples in the State of Colorado; that in that climate the apple trees . . . must be sprayed with some poisonous solution in order that apples may be grown that are fit for use at all. He admits that he did use the material . . . and that he did after these apples had matured on the trees, up until picking time, that he picked them, caused them to be picked, and shipped them to himself in Hutchinson, Kans., to his warehouse there, and he contends that at that time they were not ready for use as apples. . . . Now, it is the contention of Judge Martin . . . that he was not shipping these apples to market, that they were not ready for food; but his contention is that he was shipping them to his own warehouse and to be there kept until they reached that condition when they could be put upon the market for food, and at which time the law, as he says, would apply, but they were not at this time ready for food as an edible.

"This is a very practical world. We have to live a practical life. And amidst all of these numerous regulations we have to still live a practical life. If Judge Martin did anything in this case with these apples whereby the health of individuals might be deleteriously affected, then the apples ought to be condemned. But, while these apples were handled in interstate commerce, if they were so treated not as a food product or not with any intent of offering as a food product, if he kept them under his control so they could not go into the hands of others to be used, and affect them injuriously, then he had a right to deal with them in that practical way. For, suppose the Government should admit here that the purpose of shipping them—not only in interstate commerce—that they not only were shipped in that manner but suppose he were shipping them to feed his hogs; it was not hogs the Government was looking after when it enacted this law and the Government wouldn't have any concern whether they were good for individuals or not if they are shipped to and fed hogs, that would be the individual's business and not the Government's. But if anything is the matter, if it was not in good faith intended that these were not ready for the market and should not go on the market, then you would find the defendant guilty, if you believe the amount of poisonous products added in this case was deleterious to human health. So that is the way I am going to leave that question to you in this case, of which fact you are the sole and exclusive judges. You will take the matter and determine it in the light of the law as I have given it to you."[24]

Aside from being referred to as "the defendant," when technically he was "the claimant," Judge Martin could hardly have hoped for a more favorable charge from Judge Pollock. If not eloquently, Pollock had sedulously pried open the escape hatch Martin had provided for himself. The Wichita trial had in fact been largely contrived by Martin, who longed for a test case to "wipe out silly Federal Bureau regulations."[25] He had shipped his fruit in crates marked, "Sprayed Apples Not Wiped," and audaciously notified the Bureau of Chemistry of the shipment, then confidently waited for his day to humble the Bureau in court. Should he fail to refute the toxicologists, he planned to beat the Bureau with the technicality of his apples not being intended as marketable food.

The jurors he expected to manipulate, however, apparently saw through his scheming, and, deciding the apples had not been shipped to the warehouse in good faith, returned a verdict of adulteration on March 22.[26] Martin successfully moved for a new trial, but died before the case could be rescheduled. In April 1930, with Martin's son's consent, the apples were officially and finally decreed adulterated and ordered condemned.

Unusual for the pre-trial machinations of the claimant, the Martin case nevertheless typifies the defense of its actions that the Bureau used in other trials, as well as the points of opposition it usually encountered. The jury, however, was known as too unpredictable an institution for the Bureau to expect all prosecutions to succeed like those of Banks and Martin. The experience of the Banks trial alone was enough to suggest that if future contested produce seizures were to be sustained, some more authoritative support for the Bureau's informal arsenic tolerance might be essential; that is, rather than relying on expert witnesses to testify to the wisdom of the tolerance after the fact of seizure, it might be better to have a panel of impartial toxicologists recommend a tolerance to the Bureau before it was adopted as a guide to enforcement. The second type of tolerance should strike a jury as being more fully thought out, a product of calm deliberation outside the courtroom, not an opinion formed for one particular case, or one extracted by a government lawyer during the heat of legal debate.

Accordingly, in early December of 1926, Bureau Chief Browne submitted a memorandum to Secretary of Agriculture William Jardine: "In view of the publicity which is certain to result from the court cases now being tried with regard to the seizure of apples containing dangerous amounts of lead arsenate spray residues I wish to suggest the advisability of immediately appointing a committee of expert toxicologists who shall give the Department the best available opinion upon the limits of lead and arsenic which shall be tolerated in human foods."[27]

The Bureau's request was approved, and a committee of experts calculated to impress any challenger to Bureau policy, and any jury, was soon appointed. The Chairman was Reid Hunt, professor of pharmacology at Harvard University since 1913. Hunt's co-committeemen were Carl Alsberg, former Chief of the Bureau of Chemistry, now Director of the Food Research Institute of Stanford University; Haven Emerson, Director of the Institute of Public Health, College of Physicians and Surgeons, Columbia University; Frederick Flinn, assistant professor of physiology at Columbia, and a specialist in industrial toxicology; Arthur Loevenhart; and Carl Voegtlin, pharmacologist with the United States Public Health Service and formerly a pharmacology professor at Johns Hopkins. Anton Carlson was also appointed to the committee but was unable to attend the conference.

The "Hunt Committee" assembled for the first and last time at the January 3, 1927 "Conference on the Effects of Arsenic and Lead in food."[28] The conference's morning session consisted of a briefing supplied by the Bureau of Chemistry's Walter Campbell and by representatives from the Bureaus of Entomology and Plant Industry. All sides of the residue question were carefully examined by men with years of experience with its various aspects. "All we are asking the pharmacological experts to do," it was stressed, "is to consider the difficulty of this situation—to grow fruit and protect it from the depredation of insects, and on the other hand to protect the public from its effects, and it is a rather hard nut to crack."

The cracking of this nut—the determination of the amount of arsenical residue both safe for the consumer and practical for the farmer—was expected to be accomplished by the Hunt Committee during the afternoon of the same day. When the morning's briefing ended, the pharmacological experts were closeted in Chief Browne's office, along with any "books, records and other material which the members may desire,"[29] and left to make their decisions. The matter was not quite so hurried as the schedule might suggest, for the conference had had time to consider the problem in the preceding weeks. They were all thoroughly conversant with arsenic toxicology, being familiar not only with the long association of chronic intoxication with arsenical medicines and wallpapers, and with the Royal Commission investigation, but also with the several more recent experimental and clinical studies relating arsenic to a variety of chronic ailments.[30]

The report that emerged from their closed door deliberations, while bland in tone, was somewhat surprising in substance. "The conferees are of the opinion," the opening section ran, "that evidence of prevalence of lead and arsenic poisoning from the ingestion of fruits and vegetables sprayed with insecticides and fungicides is scanty and unconvincing, and yet, the insidious character of accumulative poisoning by these substances is known to be easily overlooked, and that the lack of evidence of prevalence of such poisoning must not be accepted as proof that such poisoning does not exist. In answer to the following . . . questions put to the conferees by the Department, they offer replies as below:

" (1-a) What are the maximum amounts of arsenic and lead which can be taken by the human subject per day without danger to the public health?

"The conferees are of the opinion that, pending more accurate information as to toxicity of arsenical spray residue when ingested with raw or cooked fruit or vegetable products, a suitable administrative rule would be that such fruit and vegetable food products be withheld from sale or use if they contain more than 3 mg. of arsenic calculated as As_2O_3 per kilo or per litre of food product.

" (1-b) The conferees are of the opinion that pending evidence more complete than any at present available as to the minimum safe daily exhibition of lead dosage without damage to health foods be not permitted in sale or use which carry the equivalent of 2 mg. or more of lead element per kilo or litre of food."

The Hunt-proposed arsenic tolerance of 3 mg. per kilo translates into 0.021 grain per pound, or slightly over twice the British or world tolerance. The recommendations for liquid foods, furthermore, is equivalent to 0.21 grain per gallon (imperial), a figure many times greater than the British tolerance and one that, it would seem, could be justified only by assuming a limited consumption of arsenical liquids (this assumption is exactly opposite to that of the Royal Commission, which, being concerned with beer rather than fruit, set stricter arsenic limits on *liquid* food).

A seeming divergence from past opinion is also apparent in the Hunt Committee's lead proposal. It may have seemed peculiar that so little has been said thus far about the danger of the lead portion of lead arsenate residues, since lead is a serious poison whose toxic properties have been known since antiquity.[31] For centuries, exposure to lead compounds had been linked to the development of chronic complaints such as constipation, colic, pallor, and nervous disturbances, including paralysis. Epidemics of lead poisoning, furthermore, were known to have resulted from consumption of lead contaminated foods,[c] so one might expect the application of a salt of lead to fruits and vegetables to excite a certain amount of alarm. But, as with arsenic, it was agricultural rather than medical scientists who evaluated the hazard of lead residues, and their concern was for acute rather than chronic intoxication. Rather large quan-

[c] The eighteenth century's Poitou colic and Devonshire colic are the most famous of a number of local epidemics recognized as the result of the contamination of wine or cider with a lead salt. A recent epidemic of gout in Alabama has also been blamed on an alcoholic beverage, moonshine, contaminated with lead (G. Ball, *Bull. Hist. Med., 45,* 401 (1971). Unlike the gout that afflicted Georgian England, this outbreak was confined to the lower end of the social scale.

tities of lead are required to produce acute poisoning; consequently, whenever lead residues were determined they were immediately dismissed as too small to be dangerous. Prior to the late 1920s, the more notorious poison, arsenic, attracted nearly all the attention of those concerned about spray residues.

The Hunt Committee toxicologists, however, were appreciative of the chronic threat posed by lead residues, and were conferring during a period when chronic plumbism from sources other than spray residues was receiving considerable scientific publicity. Just two years before the Hunt conference, a lengthy review of the state of knowledge of lead poisoning—including discussion of the routes of absorption of the metal, the symptoms produced by it, and the diagnosis and treatment of lead intoxication—was published by the Harvard toxicologist Joseph Aub. Aub stressed the insidiousness of lead poisoning. "The more severe types of intoxication by lead are very easy to recognize," he noted, "but the mild manifestations are so protean in character and develop so irregularly that differentiation between absorption and true intoxication is often nearly impossible."[32]

At the same time that the Aub article was being given wide circulation, the U.S. Public Health Service was engaged in a study of the dangers of exposure to leaded gasoline. Lead tetraethyl had been introduced as an anti-knock additive to motor fuel early in 1923, and within the next year and a half 139 cases of lead poisoning, including 13 culminating in death, were reported among workers involved with its manufacture.[33] In May 1925 the distribution and sale of tetraethyl lead gasolines was voluntarily stopped by the petroleum industry, and a conference to investigate the dangers of the product was called by the Surgeon General. Subsequently, a committee, which included Reid Hunt, was appointed to study the possible hazards of non-industrial exposure to leaded gasoline. This group concluded in *Public Health Bulletin 126*, published in 1926, that although neither service station operators nor ordinary drivers showed any signs of lead intoxication, exposure to tetraethyl lead over a period of years might result in difficult-to-diagnose chronic injury. This possibility nevertheless seemed insufficient reason to prohibit lead-containing gasolines from production or sale, and, once factory facilities were improved to eliminate accidents to petroleum workers, leaded gas reappeared on the market.

The Public Health Service study offered little indication of the quantities of lead a person might be able to tolerate over an extended time period. No less than with arsenic, this was a difficult figure to estimate, both because of the subtle nature of lead intoxication and because of the wide variation in individual susceptibilities to lead. A number of appraisals of the danger level of the metal in drinking water had been made, but these were spread over a broad range. The lowest figure to be commonly suggested was 0.50 parts per million, though experiments on lead intoxication in rats reported in 1922 had indicated that lead concentrations of 0.2 to 0.3 parts per million might be harmful to man.[34] Even more recently, another toxicologist had charged that 0.25 parts of lead per million of drinking water could injure some people, and had questioned "if we are justified in drawing any dividing line between what may be pronounced 'safe' and 'dangerous' amounts [of lead]."[35]

The Hunt Committee had been asked specifically for pronouncements of "safe" and "dangerous." While its report urged that residues of lead were more dangerous than those of arsenic, this concern was not fully reflected in its recommended lead tolerance. The proposed limit of 2 mg. per kilo equals 2 parts per million (or 0.014 grain per pound), a figure significantly higher than the values of 0.50 parts per million, and less, which had been suggested by some students of lead toxicology to be hazardous.

The apparent leniency of the Hunt recommendations can be explained. In the case of lead, it could be argued that a solid residue from fruit would not be absorbed from the gastro-intestinal tract as readily as the metal dissolved in water. Since, in addition, the average person consumes a greater quantity of water than of fruit, a more liberal tolerance for lead as spray residue would be justified. Of probably more significance, for arsenic as well as lead, was the committee's charge to take into account the farmer's need to combat insects and his ability to meet tolerance requirements. This consideration alone might account for the liberality of the Hunt tolerances. Whatever the rationale for these recommendations, the Bureau now had the expert testimony it had wanted before initiating its new domestic arsenic policy. Seven weeks after adjournment of the Hunt Conference, that policy was finally announced.

The Salt Lake City Spray Conference was convened on February 21, 1927. It was the latest in an already long line of meetings between Bureau of Chemistry personnel, other federal and state officials, and representatives of the fruit and vegetable industries. The Bureau hoped it would be the last. Repeated spray conferences had been necessary in the past because the agricultural industry was confused by the sudden imposition of residue controls and resistant to the maintenance of these controls. The Bureau now realized it had contributed to both the confusion and the resistance with its secret arsenic tolerance, and proposed to win the permanent understanding and cooperation of agriculture by openly explaining its new residue policy. As if beginning a new year, the Bureau went to Salt Lake City acknowledging its past bad habits and resolving not to continue them.

The Bureau spokesman was Walter Campbell. Addressing a congregation of agricultural officials and scientists from the western states, Campbell outlined the history of residue regulation since 1919, emphasizing the patience and sympathy the Bureau had extended to the agricultural industry during those difficult years. But it should be realized, he continued, that the period had been trying for the Bureau as well. By regularly placing its duty to protect the public health subordinate to its desire to spare agriculture excessive financial loss, the Bureau had put itself in a position where it "might be severely castigated by self-constituted public health crusaders if its leniency became known."[36] This leniency could not, would not, go on. Cleaning methods were available that, if conscientiously applied, could remove all or most of the arsenic from sprayed produce. Therefore, "There is no question but what the Department must insist that the crop of 1928 shall meet the 0.01 grain tolerance."[37]

Two questions are at once provoked by this announcement. Why was the British arsenic tolerance selected by the Bureau instead of the more liberal recommendation of its own hand-picked Hunt Committee? An intriguing possibility is that the Bureau had definitely decided to adopt the British tolerance before the Hunt Committee was even appointed. It was the tolerance honored throughout Europe. Toxicological opinions solicited by the Bureau in 1926, it will be recalled, favored the 0.01 grain per pound limit. In contacts and correspondence with the produce industry, Bureau personnel themselves had for some time been defending the validity of the British arsenic tolerance and promising eventually to employ it for domestic operations. The Hunt Committee might thus have been appointed primarily to bestow an appearance of ultimate authority on a decision that the Bureau had already reached. The agricultural industry, after all, could be expected to comply more readily with a tolerance believed to bear the imprimatur of American experts than with one suggested a quarter-century before by a group of foreign scientists. When the Hunt Committee failed to rubber-stamp the British tolerance, its own more lenient tolerance had to be ignored. If the Bureau had proposed 0.021 grain per pound as the permanent arsenic tolerance after having previously insisted that 0.01 was the maximum safe figure, the result might have been more than temporary embarrassment. Farmers could reason that a Bureau which had already erred so badly, by more than one hundred percent, might still be wide of the mark. Residues might be taken more lightly than ever by agriculturalists if the true Hunt recommendation were disclosed. Early in his remarks at Salt Lake City, Campbell went so far as to imply that 0.01 was the tolerance recently recommended by a government-appointed "committee of physiological experts."[38]

A less conjectural explanation for the Bureau's insistence on the British rather than the Hunt arsenic tolerance is that the lower figure was necessary for the control of lead residues. The Bureau had been deeply impressed by the Hunt Committee's warning that lead "presents an even more serious menace than does . . . arsenic,"[39] but found itself incapable of dealing directly with this menace. The state of analytical chemistry in the 1920s was such that the accurate determination of small quantities of lead in organic matter required the better part of three days. By the time Bureau chemists could ascertain that a sample of fruit carried excessive amounts of lead and recommend seizure action, the fruit might be shipped, marketed, sold, and consumed. Until a rapid method of lead analysis was developed in the 1930s, lead residues had to be indirectly monitored through arsenic. The chemical formula of lead arsenate $(PbHAsO_4)$ suggests a weight ratio of lead to arsenic of two to one.[d] An arsenic tolerance of 0.01 grain per pound would thus be equivalent to a lead tolerance of 0.02. This is a considerably higher allowance for lead than the Hunt Committee

[d] Actually, as was later discovered, weathering tends to remove arsenic more easily than lead, so that lead to arsenic ratios in residues are often significantly greater than two to one; see *Journal of Economic Entomology, 31*, 594-597 (1938).

recommendation of 0.014 grain per pound, but it could be made no lower, for an arsenic tolerance below 0.01 simply could not be enforced. In 1927, the British tolerance for arsenic was, for the time being, the most effective stratagem for minimizing lead residues. It was as much for this reason as for its concern about arsenic that the Bureau settled on 0.01 grain per pound as its domestic arsenic tolerance.

The second question raised by Campbell's announcement of this tolerance for the 1928 season is that of the form of regulation intended for 1927. Although past leniency had to stop, it might have to be brought to a halt gradually. Campbell knew that "It is too much to expect . . . that the industry of the West will so generally adopt approved cleaning processes for 1927 as to meet this requirement for the current year's crop." The Bureau, on the other hand, "cannot operate during 1927 on the same extremely liberal tolerance it employed last year, for such a policy would be a tacit admission that no progress whatever has been made in the control of the situation. It would only be human nature for the industry to postpone the adoption of necessary corrective steps for another year if it were given to expect that no better cleansing would be required in 1927 than was expected in 1926."

The only solution was compromise. "It is fair to expect that the industry will meet a mean between last year's extremely liberal informal tolerance and that which will be required next year."[40] The exact level of the interim tolerance was then thrown open for debate by the assembly, but the discussion was brief and friendly and by the afternoon unanimous consent had been given to resolutions that: "A tolerance of 0.01 grain of arsenic . . . per pound of fruit be in effect for the fruit crop of 1928. . . . That in recognition of the difficulties of installing the necessary facilities for reaching this tolerance in all cases for the 1927 crop, seizure or prosecutions be not instituted on that year's crop, unless the amount of arsenic . . . found shall exceed 0.025 grain per pound."[41]

It was further resolved that these tolerance proposals be submitted for approval to the other section of the spray conference, the representatives of the fruit industry who had been meeting in an adjoining room. Ordinarily, their approval of such resolutions would have been grudging at best, but the fruit men had been put in an agreeable frame of mind by their speaker, the International Apple Shippers' Association's R. G. Phillips. Phillips' speech had been lengthy, but his message was the simplest: the industry was courting financial disaster by not cleaning its products more thoroughly. Continued conflict with the law would result in adverse publicity leading to embargoes abroad and public panic at home. Think, Phillips exhorted his audience, of the "chaos and bankruptcy [which] would reign in the industry in this country," and of the "emotional people . . . in this country. It takes but little to arouse them to fever pitch. It would take only the turn of a hand to excite thousands of consumers . . . to the most violent action and demands."[42] Thus frightened into enlightenment, the industrial delegates did not hesitate to adopt the other section's resolutions as their own. Both groups departed from Salt Lake City eager to spread the word of the new residue policy throughout the fruit industry, and confident that a new understanding between farmers and the law had been achieved.

The spirit of Salt Lake quickly soured. On July 1 the Food and Drug Administration replaced the Bureau of Chemistry, and within the month the Chief of the FDA's Denver Station was complaining that practically none of the fruit growers in his area "had any knowledge of the epoch-making spray residue conference held last February."[43] Even when aware of the "epoch-making" conference, few farmers approved of it. Another FDA Station Chief reported that "almost constantly throughout the past season, both at Wenatchee and at Yakima [the two major fruit-growing centers in Washington], growers individually and also the officers of the different farm organizations . . . frequently stated that the growers were not properly and adequately represented" at Salt Lake City.[44] What the growers and officers meant was that their representatives had made promises in Salt Lake City that could not be kept without sacrifices of time, effort, and money greater than had been anticipated. The removal of residual arsenic, even to a level of only 0.025 grain per pound, was not without complications.

Ever since the first seizure of pears in Boston, the owners of confiscated produce had been given opportunity to repair it to marketable condition, but more often than not the owner's efforts failed and his produce had to be destroyed. One of the topics of most concern at the first spray conference in 1919, as well as at those which followed, was that of techniques of residue removal. During the early years, wiping of produce was most often suggested, and federal and state scientists cooperated with members of the agricultural industry to develop machinery to make wiping economical.[45] Several models were introduced. One of the simpler machines carried fruit on a conveyor belt beneath revolving brushes that rubbed arsenical residues away. An advance on this model had brushes installed on either side of, as well as above, the fruit conveyor, while yet another machine operated by a similar design but with cloth disks instead of brushes. Even the most sophisticated dry-cleaning methods were unable consistently to remove more than about thirty percent of the residue from heavily sprayed fruit, however, and, in the attempt, fruit was often bruised by the mechanical wipers. On occasion, fruit even left the machines carrying more arsenic than it had had upon entering: the brushes and wipers in the machines could become so heavily laden with arsenic removed from fruit that unless they were replaced with clean parts at regular intervals, fruit would begin picking up arsenic from the wipers rather than vice versa.

The possibility of dissolving residues in dilute acid had also been raised at the 1919 spray conference, and soon afterwards the Department of Agriculture's Bureau of Plant Industry began investigations directed toward the development of fruit-washing machines. The work of the chemist Arthur Henry was particularly important for the construction, by the mid-1920s, of effective acid-wash residue removal machines, and Henry himself was active in demonstrating the machines to growers. When the Bureau of Chemistry learned that a private firm was also developing a fruit washer and planned to patent the process in order to receive royalties from fruit growers, the Bureau went to considerable lengths to secure a public-service patent for Henry's process and thereby save farmers the expense of royalties. Henry's and other washers, some using alkaline washes in place of or in addition to the acid one, were being employed throughout western fruit regions by the late 1920s. The most commonly used washing machine passed fruit on a conveyor either through a flood of washing solution flowing in the opposite direction, or under a spray of washing solution. Fruit was also cleaned by floating (with periodic immersion) in the cleaning solution, or even by being packed in crates and submerged in washing solution for several minutes.

Such a prosaic operation would hardly seem capable of stirring the passions, but residue removal was the most emotionally charged of issues for many farmers from the mid-1920s through the decade of the 1930s. It engendered among orchardists a remarkable amount of ill feeling toward their government, and inclined more than a few toward paranoia. The bitterness and suspicion were much more pronounced in the West than in other areas of the country, for western growers were much more seriously affected by residue regulation. The heavier insect infestations of the West forced them to spray more frequently, and the drier climate was less efficient at removing residues by weathering. Most of the FDA's seizures, consequently, were of shipments of western produce, and westerners bridled at this treatment. The peak of anti-government sentiment was

166

reached in the greatest apple-producing region of the country, the state of Washington's Yakima Valley.

> *There's a vale of peace and plenty where the big*
> *   red apples grow;*
> *Where the luscious peaches ripen and the clustered*
> *   grapes hang low;*
> *Where the hand of man has made the desert*
> *   blossom fair to see;*
> *Where the Yakima flows onward in its journey*
> *   to the sea.*[46]

The orchardists of the Yakima Valley were proud of their land and of what they had done with it. The area's first settlers, arriving in the mid-1800s, had found it an arid region with an ashy soil, a near-desert of little promise. "And yet with the naked eye could be seen mountain peaks with their eternal snows which vernal days turned into tumbling waterfalls, ever-flowing creeks and rivers brimful of harvest-giving water."[47] The channelling of the water from the "eternal snows" of the Cascades into irrigation canals transformed the Yakima Valley into a land of plenty. Commercial apple orchards began operation in the 1880s, and in 1900 shipments of Washington apples to other states were begun. By the 1920s, thousands of carloads of apples, pears, peaches, and other fruits left Yakima every season for points all over the world. The world's champion winesap tree grew in the Yakima Valley. Yakima fruit was selected to help provision Admiral Byrd's expedition to Antarctica. It was no wonder that while other regions of the country exaggerated and distorted the facts when boosting their local attractions, the Yakima Chamber of Commerce made its motto, "The Truth About Yakima is Good Enough."

The marvelous progress of the Yakima Valley had been hard-earned, achieved only by the investment of millions of man-hours in the construction of irrigation systems, the planting of fruit trees, and the pruning and spraying of the trees. It was progress fraught with uncertainty all the way.

Frost, insects, and plant diseases threatened the orchardist every spring and summer. The months after the harvest were filled with worry that the fluctuations of supply and demand would cause price drops that would erode, even destroy, the season's anticipated profits. For Yakima orchardists, and other western growers, there was the added burden of inflated railroad rates. A considerable portion of the expenditures of the western fruit industry went for the transportation of its product to the consuming centers of the East. Unseemly from the beginning, rail rates had been rising steadily since 1913. By the mid-1920s, western fruit districts were in a state of constant irritation from the gouging of the railroads.

Already feeling hard put to preserve some return from their long and arduous labors, western fruit growers were in no mood to be told that spray residues would have to be removed from fruit before it could be marketed. The announcement was taken with surprising equanimity for a period, but as hand wiping of fruit, then machine wiping, were found inadequate to meet the government's standards, the rebelliousness of farmers boiled over. Even the introduction of washing machines that effectively removed residue only made matters worse in one sense, for while the machines' developers cheerfully announced that their operation would add only one to five cents per bushel to the price of fruits, these costs seemed anything but negligible to the men who had to pay them. A northwest orchardist of average means calculated that "at five cents per bushel, [that] would be $55.00 per car; on fifty cars that would catch me for over $1,500.00."[48] Coming as it did on top of all the other troubles and expenses that beset the fruit grower, residue removal seemed certain to tip the unsteady seesaw between profit and loss in the direction of financial ruin. Even before the Salt Lake City conference met, a Yakima area orchardist wrote his Senator:

"I suppose you are aware that the Chemical Department at Washington, D.C. [the Bureau of Chemistry] has ruled

168

the N.W. apple business homeless, by the ruling on arsenate of lead residues. This will make homeless, nothing to eat, nothing to clothe, 100,000 families at least. Can you comprehend the enormity of this ruling, yet no one of our Congressmen have endeavored to condemn such a ruling as far as I know. Is the Chemical Department prepared to feed this half a million souls, beside clothe and make them homes. Besides the ruin of much business, possibly 100 cars out of a possible 40,000 carload crop might be able to pass this chemical test. Please do your best to get Washington, D.C. into action. Whatever is done must be done speedily or we are ruined. (P.S.) The whole district is up in indignation about it."[49]

The predictions of calamity, fortunately, failed to come true, but they continued to be made nonetheless. Overwrought accounts of the hardships imposed by spray residue regulation were to be regularly delivered to Congressmen from western states for years to come. They were symptomatic of the western fruit producer's conviction that he was being persecuted by his own government. He could not avoid noticing that nearly all the shipments of sprayed produce that had been seized by the FDA had been grown in western states. Why, many asked, are "we, the Western Fruit Growers, . . . penalized by being compelled to wash the spray from our apples as they come off the tree and ship the same to both domestic and foreign markets?"[50] The logical answer that eastern states had fewer insects and a wetter climate, and thus produced fruit with lower residues, did not ring true to western growers, already annoyed by the eastern industry's competitive advantage of being much closer to the large marketing centers, and thus having much lower transportation costs. Regional discrimination seemed a more plausible explanation. By the summer of 1927, FDA personnel in the Yakima district were aware that growers there "have come to believe that the whole spray residue proposition and washing is a scheme put over on them by the Government to put the Northwestern growers out of business and that the manufacturers who are making the [washing] machines are in on the conspiracy."[51] Just why, and with whom, the government would conspire against the western states was never very clear, but the suspicion that it was doing so was hardly confined to Yakima. In their impotent rage against their oppressors, the growers of some areas pondered the feasibility of suing railroads that co-operated with FDA inspectors wanting to examine fruit shipments.[52] Others intimated to their congressional representatives that the western agricultural community was being driven toward Bolshevist revolution by the tyrannical residue regulations.[53]

Yet as misinformed and unfair and discriminatory as the government's policy seemed to be, it was the law, and most farmers prided themselves on being law-abiding citizens. The majority of growers thus tried to comply with the residue regulations, though their efforts were rarely more than halfhearted since few could convince themselves that sprayed produce was really dangerous. The views of the dozens of Yakima Valley orchardists who wrote in complaint to the Department of Agriculture were succinctly represented by the man who observed: "We don't know what it is all about, but presume some professor heard that someone said another had learned someone somewhere got too much poison. The writer has lived here twenty-three years where men, women, children, cows, calves, horses, colts and pigs eat these apples; and has never heard of a case of illness due to this nor of anyone else who has heard of a case. That's that."[54]

The voice of experience spoke louder to the farmer than did the opinions of university professors who had never eaten sprayed fruit direct from the tree, but who for theoretical reasons considered such fruit dangerous. It was simply "impossible to agree with the gentlemen who, all at once, have discovered that this [sprayed fruit] is a poisonous and adulterated product."[55] To agree with this untested opinion would be to abandon not only common sense, but sound economic philosophy. "Personally," one orchardist informed his senator, "I never have heard of any person being injuriously affected from the arsenic on account of eating apples. While some of the professors and scientists tell us that the present arsenic content of sprayed fruit may be injurious to certain persons, I feel that we are permitting this technical, scientific opinion to carry us beyond the practical application of business principles, and are thereby irreparably injuring an industry which is in need of our help."[56]

Despite the apparent strength of the case against residue regulation, as silly and pointless as the washing of fruit seemed to be, so long as the FDA's bureaucrats believed in its necessity, residue removal would have to be practiced by farmers. They would practice it resentfully, but they could also derive a certain satisfaction from rising to meet this latest adversity. Thus while the Yakima Fruit Growers' Association snickered at being "compelled to pass our entire crop . . . through a test tube,"[57] it also promptly responded to the news of the Salt Lake City Spray Conference with the assurance that the British tolerance could be met by 1928 "without undue difficulty or expense."[58] The announcement that Yakima orchardists " 'can do it this year (1927) if necessary,' brought an outburst of applause from the members."[59]

Residue removal was more easily said than done, and as the summer of 1928 approached, the FDA offices were flooded with mail from all the western fruit-producing regions, requesting postponement of the British tolerance. The letter of a Yakima district orchardist was typical: "I feel the Northwest apple industry has made a most commendable effort to comply with the spray residue regulations, and has succeeded very well but owing to difficulties encountered should have another year before establishment of the 0.01 limit."[60]

The difficulties encountered were the result of farmers' spraying too often and too late in the season with mixtures that employed oil or casein as spreaders. Fruit so treated could not be consistently cleaned by the best washing machines. FDA Director Campbell received so many complaints that the British tolerance was impractical for the 1928 season that on February 24 he mailed the following questionnaire to a number of prominent figures in the fruit industry:

"While the Salt Lake City conference agreement contemplated the adoption of the world tolerance of 0.01 grain arsenic . . . per pound of fruit for 1928, and while decided progress has been made in the reduction of the spray residue heretofore found, the Department has received a large number of statements from most if not all of the producing sections to the effect that because of the lack of adequate facilities, if for no other reason, it will be impossible to meet the world tolerance universally during the coming season.

"In conformity with the moderate administrative policy which the Department has sought to maintain in the past in connection with this problem, it is desirous of imposing no requirements during the 1928 season which are impossible to meet. If in the opinion of medical experts 0.01 grain of arsenic . . . per pound represents the safe limit of spray residue, it is imperative that this limit be reached and this should be done as soon as possible. In order to permit the Department to determine what stand is possible and right for the 1928 season. . . . I am writing to ask for a frank discussion from you of your observations during the past season and your opinion as to what course should be adopted for the coming season.

"(1) Do you believe the Salt Lake City conference agreement should be carried out literally and the 0.01 tolerance be put into effect without exception for 1928? If so, why?

"(2) If not, what figure do you think should be put into effect for 1928, and why?

"(3) Do you consider that the 0.01 tolerance should and can be met in 1929?

"(4) Have you any reason for believing that the acid wash process cannot be used successfully on a commercial scale?

"A prompt and comprehensive reply and discussion will be greatly appreciated. An addressed franked envelope requiring no postage is enclosed for your reply."[61]

The replies predictably answered question (1) "no" and question (3) "yes," and the FDA was left little choice but to modify its Salt Lake City schedule. For 1928, it was finally agreed, the administration's tolerance for arsenic would be 0.02 grain per pound, and the world tolerance of 0.01 would be reserved for the 1929 season. One year later, amid grumblings that the world tolerance could not be met by 1929 after all, the FDA polled the fruit industry again, and again revised its schedule. The tolerance for 1929 was reset at 0.017 and 0.010 postponed until 1930. In 1930, the arsenic tolerance actually employed was 0.015; in 1931, it was 0.012; and not until 1932 was the long-promised world tolerance at last adopted.[62]

The formulation of public health policy by the economic group responsible for the hazard to be regulated has associated with it certain obvious dangers, but in the case of spray-residue control, this undesirable situation was largely unavoidable. For every letter orchardists sent to the FDA, at least one more belligerent one was mailed to a congressman. And, just as the Supreme Court follows the election returns, the Food and Drug Administration had to be responsive to the mood of Congress. Its operations were financed by a congressional appropriation, and an angered Congress could easily restrict FDA activity. It might even eliminate residue control altogether, as very nearly came to pass. The apple growers of Colorado were as upset as any group by residue regulation, and their complaints were repeated loud enough and long enough to rouse their Senator W. C. Waterman to action. In 1928 the Senator proposed, as an amendment to the McNary-Haugen Farm Relief Bill, that the provisions of the Food and Drugs Act be declared inapplicable to "any fresh or natural fruit in the condition when severed from the tree, vine, or bush upon which it was grown." Waterman justified the amendment on the grounds that "it will be extremely beneficial to the fruit growers of the West and relieve them from a burden under which they have been suffering now for the last five or six years."[63] Harvey Wiley, still stumping for stricter enforcement of his pure food law, recognized the Waterman proposal as a potential "mutilation" of the Food and Drugs Act,[64] but too few congressmen were disturbed by this possibility to prevent the amendment from riding the Farm Relief Bill through first the Senate and then the House of Representatives. Waterman's scheme finally was thwarted only incidentally, when the McNary-Haugen Bill was vetoed by President Coolidge, who regarded its attempts at price fixing of agricultural commodities unconstitutional.[65]

The close call of the Waterman amendment heightened the FDA's awareness that the agricultural industry would not be driven too rapidly into compliance with strict residue regulations. There were a carrot and a stick to which the industry would respond, however. The promise of shipments unimpeded by FDA inspections drew fruit growers on toward attainment of the world tolerance, while they were goaded ever faster in that direction by the fear of publicization of the residue issue and the public agitation that would ensue. The FDA's commitment to residue secrecy had been reaffirmed at the Salt Lake City conference, but the fruit industry knew that no secret can be kept forever, and that it would be serving its own best interests if it could lower residues to unquestionably safe levels before the existence of residues became public knowledge. The situation was summarized nicely by the Chief of the FDA's St. Louis Station in a 1929 address to the American Pomological Society:

"What do you suppose would happen if the general public became acquainted with the fact that apples were likely to be contaminated with arsenic? . . . So far, we have not

174

given the matter any publicity, and the public as a whole has no general knowledge on the subject. We will be in a more enviable position, when all apples are satisfactorily cleaned to say to the public, if it gets curious, that the apples produced in this country can be eaten with perfect safety. If they were to become curious today, we would probably have to admit that we are perhaps not doing everything possible to remove excess arsenic from our fruit."[66]

What these comments failed to convey was that the public was already becoming curious, and would soon be positively inquisitorial.

*At first sight it seems somewhat perplexing
that so powerful a poison [arsenic] should so
universally be found in persons leading
commonplace lives in an ordinary environment,
but a little investigation reveals the fact that
the world we inhabit is permeated by this
subtle poison so that the possibilities for its
accidental absorption are countless. . . . It
is an uncanny thought to realize that this
lurking poison is everywhere about us, ready
to gain unsuspected entrance to our bodies
from the food we eat, the water we drink, and
other beverages we may take to cheer us, the
clothes we wear, and even the air we breathe.*

—Karl Vogel, *American Journal of the
Medical Sciences, 176,* 215 (1928)

Regulatory Publicity 6

THE CHEMICAL ARTIFICIALITY of the modern industrial environment, and the dangers this portended, was a subject that
began to attract considerable medical comment during the
1920s. The voice of Dr. Vogel was but one of many raised
in protest during these years against the growing contamination of man's surroundings with toxic metals. Arsenic
and lead drew most of this fire, both because they were
among the most dangerous metallic poisons, and because
each was being dumped into the environment from so many
sources. Although few wallpapers still contained arsenic,
Vogel did find the poison generally present in smelter exhaust gases, in the coal smoke that befouled American cities,

in fruits and other foods bleached with contaminated sulfur, in a variety of drugs, and in an array of common household items that ranged from weed-killers and fly-papers to paints, playing cards, artificial flowers, candy wrappers, and even to leather hatbands and hair tonics. The American food supply, he pointed out, was largely poisoned by lead arsenate and other arsenical insecticides, though the lead compound was considered the most serious offender because it contained a second poison. Insecticidal lead, he might have added, was only a single item on an impressive list of sources of lead pollution. A survey of the lead toxicology literature of the 1920s turns up criticisms of paints, canned goods, metal containers, cooking utensils, piped water supplies, and automobile and smelter exhaust gases for their contributions of lead to the human environment. Distribution of lead and arsenic was so complete that all members of industrialized populations carried at least traces of the metals in their tissues, and there was some serious scientific consideration being given the notion that these were normal physiological constituents.

From the late 1920s through the following decade, none of the new sources of lead or arsenic was decried more frequently by medical scientists than the insecticide lead arsenate. The phenomenon is quite interesting, and at first surprising, in view of the medical profession's general disregard of arsenical insecticides in the years preceding the 1920s. The sudden appearance of scientific resistance to insecticides would seem the result of a confluence of trends: of the rising uneasiness about arsenic and lead contamination generally, which moved scientists to inquire after all possible sources of pollution; of the Bureau of Chemistry's and the FDA's increasing reliance on physiological experts as consultants and as trial witnesses; and of an increase in the number of clinical conditions associated with arsenic or lead in which sprayed produce was identified as a likely source of the poison. All these developments operated to draw the scientific community of the late 1920s toward a

177

realization of the existence of spray residues and of the danger these posed to public health. A sampling of the warnings that followed will indicate the nature and intensity of medical anxiety over residues.

The only conspicuous critic prior to the 1920s had been the physiologist Anton Carlson. His fear that arsenic and lead residues might cause "lowered resistance to disease, lessened efficiency and shortening of life" was noted in Chapter 3. In 1928, writing in the widely circulated journal, *Science*, Carlson again cautioned that, "We may not . . . consume enough lead and arsenic in our fruit to produce acute poisoning and tissue injury, but who is there to say that this slow assimilation of metallic poisons brought about by modern industry is without danger and ultimate injury?"[1]

In 1929, the *American Journal of Public Health* carried an article critical of spray residues and suggesting that the world tolerance for arsenic should be much reduced.[2] That same year, the New York physicians C. N. Myers and Binford Throne initiated what was to become for them a virtual crusade against spray residues. They had for some time suspected that accidentally absorbed arsenic was an etiological factor in the many cases of eczema they encountered at their New York Skin and Cancer Hospital.[3] In 1929 they charged not only arsenic, but also lead—the two metals "associated with the great menace of the decade—insecticides" with "deleterious results [which] do not appear even as a chronic effect. . . . The *latent* effects of metals attack more individuals than those with the chronic and acute symptoms."[4] Such latent damage was made all the more likely, they argued further, by the facts that lead arsenate was only one of many sources of lead and arsenic in the environment and that the use of lead arsenate was increasing drastically. From eleven and one-half million pounds in 1919, the annual application of the insecticide in the United States alone had risen to twenty-nine million pounds by 1929.[5] Calcium arsenate was also being used at a level of twenty-nine million pounds per year in the cotton states by the late 1920s, while it had been applied at a rate of only three million pounds in 1919. Serious health problems must inevitably result from such a trend, Myers and Throne urged, and they would be problems that could never be remedied by present FDA policy. "It may seem to present insurmountable difficulties to those who desire to contribute to our artificial modes of living," they concluded, "but pure, uncontaminated foods and drugs must be obtained. The fixing of any limit of tolerance is and only can be relative. Standards should aim at no metal at all."[6]

A year later, Myers and Throne presented evidence linking arsenic to eczema in children, and placed considerable emphasis on spray residue as a source of arsenical contamination.[7] Dr. A. F. Kraetzer announced the same month, in the pages of the *Journal of the American Medical Association*, that a number of cases of eczema in infants had apparently been produced by arsenic in their mothers' milk. Several sources of contamination of the mothers' systems were suggested by further investigation, but "the enormous use of arsenical insecticides makes fruits and vegetables a potent source of poisoning."[8]

Shelden suggested in 1932 that at least seven of the many cases of neuritis he had recently treated were chronic conditions produced by lead arsenate,[9] and the next year Myers and Throne published still another condemnation of arsenical insecticides.[10] The dermatologists Ayres and Anderson described a number of cases of skin disease, including cancer, in 1934, and observed that "as long as anyone can buy unlimited quantities of . . . lead arsenate such occurrences may be expected."[11] Another Myers and Throne criticism was issued in 1935,[12] the American Medical Association's *Journal* editorialized that "spray residues must constitute an important menace to the public health,"[13] and the American Public Health Association held a symposium on the residue problem. One of the symposium speakers related spray residues to "food poisoning outbreaks which crop up overnight, and very often leave the health officer without a suitable answer or explanation."[14]

A. B. Cannon, associate professor of dermatology at Columbia University, published in 1936 a review article on "chronic arsenical poisoning" in which he blamed spray residues for much chronic illness often dismissed by physicians as "ptomaine" or some personal idiosyncrasy. To add substance to his point, Cannon formulated a typical breakfast, lunch, and dinner menu, purchased the food for it at local markets, then had average portions of each item analyzed for arsenic. "The total amount of metallic arsenic for the three sample meals," he found, "would be more than ten times the amount contained in a maximum daily dose of Fowler's solution as prescribed by the present writer."[15]

The following year, Stanford pharmacology professor P. J. Hanzlik warned of "the slowly developing and continuing effects of tiny quantities [of arsenic and lead] to which vast populations are exposed, day in and day out, under the modern conditions of our existence." From analyses of marketed apples, he estimated that "for the effects of lead, one whole contaminated apple, eaten daily for weeks or months would be hazardous to health; for effects of arsenic, three or four apples daily."[16]

In 1937, also, there appeared the ultimate sign that spray residues had become a problem to be seriously considered: a Ph.D. dissertation was inspired by the realization that "the extensive use of arsenic compounds in insect control has made the toxicity of orally administered arsenic a subject of considerable interest." Sister Mary McNicholas, of Catholic University, conducted "A Study of Some Effects of Ingested Arsenious Oxide" (later published in book form) that revealed that several forms of tissue degeneration were produced in rats by the consumption of small quantities of arsenic over an extended period. She concluded that these findings seem "pertinent today because of its possible application to man who is exposed daily to similar acute and chronic intoxications in consequence of the extensive use of arsenic sprays for insecticidal purposes. It is obvious from our experimental work that the cumulative effect of arsenic is injurious to a typical and representative mammal, the albino rat; we can, therefore, conclude that similar morbid conditions will be manifested in human tissue exposed to the same or similar injurious agencies."[17]

Sister McNicholas' argument—that if arsenic administered experimentally to rats produces injury, spray residue must injure people—would have been challenged by agriculturalists as highly hypothetical. Much of the rest of the medical literature incriminating insecticide sprays, imposing though it was in volume, would also have seemed unsubstantiated theorizing to practical-minded farmers. But the case against spray residues did not rest solely on questionable allegations such as the Myers and Throne subchronic, "latent" effects. Doctors could point to actual cases of acute poisoning, some resulting in death, which could be attributed only to oversprayed produce. Professor Cannon, for instance, in his 1936 article, cited several cases of acute illness occurring the previous year that had undoubtedly been caused by arsenical apples. Among other reports in the literature were the cases of the Illinois ladies poisoned by sprayed asparagus served at their country club,[18] and of the Montana residents made ill by apples and pears. When Montana health officials began to seize oversprayed fruit, local fruit dealers objected, "but after the death of a fifteen-year-old Billings girl in the spring of 1933, attributed to arsenical poisoning from eating spray-residue-bearing fruit, no more complaints were received about our campaign against such fruit."[19]

Perhaps the worst outbreak of arsenical poisoning occurred but a few months after the Montana incidents: "During the summer of 1933 physicians of Los Angeles and vicinity were called to attend a surprisingly large number of persons suffering from acute gastroenteritis, chiefly in the form of vomiting, diarrhea, and abdominal cramps. In some cases there was also fever up to 102 degrees F; in several there were bloody urine, stools, and/or vomitus, cold sweats, thready pulse, and great prostration. Thanks to an active local Health Department, those cases were promptly investigated. Thirty in one locality were traced to a Sunday meal in which cole slaw was served to all. Cabbage secured from the same lot from which the slaw was made, showed by analysis .354 grains of arsenic trioxide per pound, or more than thirty-five times the limit of safety fixed by the United States Government. One child from a neighboring town had been stricken with vomiting, cramps, diarrhea, and prostration after eating an unwashed and unpeeled pear. Three other children who had eaten washed pears of the same lot were not ill. Analysis of unwashed pears from this lot showed .538 grains of arsenic trioxide per pound, or nearly fifty-four times the limit of safety. In addition to cabbage and pears, arsenic was found also on celery, broccoli, and spinach from various markets in the vicinity. One patient was seized with severe vomiting which lasted for several hours, after eating a single stalk of celery. An elderly woman whose supper had consisted entirely of boiled spinach was awakened early the next morning by vomiting, diarrhea and cold sweats. Urine, feces and vomitus from these and other cases . . . yielded arsenic in significant amounts."[20]

Surprisingly, there remained scientists who defended the safety of spray residues. The name of T. J. Talbert, in particular, began to appear frequently in objections sent to the FDA by farmers during the 1930s. Talbert was a horticulturalist at the University of Missouri Agricultural Experiment Station who published a series of papers (1930-1934) purporting to analyze the question, "How poisonous is spray residue?"[21] These papers were circulated widely among agriculturalists and came to be cited repeatedly as the scientific refutation of the government's position on residue toxicity. It was Talbert's considered opinion that "there is little likelihood of a human consuming as spray residue on apples, sprayed and handled in the usual manner, enough arsenic either at one time or over an extended period to be injurious."[22] The most interesting aspect of this conclusion is that it seems to have been reached almost in spite of, rather than because of, the evidence that Talbert was considering.

The "wild stories" of residue poisonings, he began his analysis, were false. "The ill effects produced have usually been due to over-eating or eating green and immature fruits."[23] One can be certain of this, he continued, because Styrian peasants had for years eaten much larger amounts of arsenic than ever occur on fruit. Many Americans, moreover, had also been consuming considerable quantities of arsenic throughout their lives, ingesting it with seafood, and had not been injured. This reference was to the recent discovery[24] that most species of shellfish have a natural arsenic content many times higher than the British tolerance.[a] U.S. Bureau of Fisheries studies had soon indicated that the arsenic of seafood occurs in an organic form that is much less toxic than inorganic arsenic,[25] meaning that the question of the safety of shellfish was only tenuously related to that of the safety of spray residues. Talbert nevertheless discussed the Bureau of Fisheries findings in an apparent attempt to imply that they supported his position. As further support, he summoned the nearly twenty-year-old guinea pig experiments of W. C. O'Kane, experiments analyzed in Chapter 3 and found deficient in their evaluation of the chronic intoxication hazard. Talbert also presented results from his own experiments, though it is not clear why since these had shown rats, animals suspected of being more resistant to poisoning than human beings, to

[a] This was a revival of the argument used three decades before by Harvey Wiley's opponents, that poisons occurring naturally in some foods cannot be prohibited as additives to other foods. Even Swann Harding, a tireless writer in support of strict food regulation, had to agree that, "It would be very difficult to convict Nature in court of the charge of poisoning human beings by putting excessive quantities of arsenic in haddock or in prawns." *Sci. Am.*, *149*, 198 (1933).

be very definitely injured by several months of feeding with concentrations of lead arsenate only four times the British tolerance.

The logic of his concluding argument was even more enigmatic. Many water supplies contain lead, he noted, and from some of these the drinking of as little as ten glasses of water could introduce as much lead into the body as the eating of an apple. This information was intended to refute "some excitable persons who charge the eating of sprayed apples causes a dangerous accumulation of lead in the human system,"[26] presumably by suggesting that if ten glasses of water are safe, then so are sprayed apples. Talbert did not claim that the ten glasses would, in fact, be safe nor did he mention that many toxicologists had published warnings that a daily consumption of much less water-borne lead than that dissolved in Talbert's glasses could produce chronic intoxication.[27]

The FDA was fully appreciative of the flaws in Talbert's arguments,[28] and of the indefensibility of his conclusion that residues were harmless. The Administration's files already contained a number of reports of acute poisonings from residues, including that of a woman who had tested the sprayed apples she believed had made her so ill by feeding several to her goat. The goat died.[29] All these cases, that of the goat included, were nevertheless regarded as exceptional. Most produce did not carry sufficient arsenic and lead to cause acute illness, and it was chronic injury that remained the FDA's primary concern. But though the FDA files were confidential, and it continued to honor its pledge to the agricultural industry not to publicize the residue hazard, knowledge of spray residues was spreading beyond the pages of medical journals. The brief newspaper notices of British objections to American apples and of occasional local actions against sprayed produce had produced some public uneasiness by the end of the 1920s, as witnessed by several letters received by the FDA. The extremes of this early public reaction were represented, on the one hand, by the Brooklyn man who resignedly asked, "Will you please state how a man who eats many apples can avoid the danger of arsenic poisoning. I am afraid I am a victim";[30] and on the other by the self-appointed pure food crusader from Los Angeles who declared that "ARSENICAL FOOD-POISONING IS A NATIONAL MENACE," and proposed new legislation to rid the nation's food supply of the poison.[31] What these early complaints against residues had in poignancy or in fervor, however, they lacked in volume. The number of citizens alarmed by the excesses of spraying was for some time miniscule beside the number of agriculturalists angered by restrictions on spraying, and public opposition to insecticide residues did not become a factor to be reckoned with until well into the 1930s. The medical literature was too technical and inaccessible, the newspaper articles too infrequent and muted for either source of information to activate any large-scale public protest. As had occurred with the pure food and drug agitation prior to 1906, the spark to set off citizen indignation had to be provided by articles and books of general distribution, which presented the facts in a form comprehensible to the layman and accompanied by commentary designed to raise his ire. It was the revival of muckraking journalism that at last created a public consciousness of the dangers of spray residues.

"Muckraking" was the term employed by Theodore Roosevelt to characterize the preoccupation of many journalists of his Progressive Era with dredging up and exposing to public view all the filth and ordure buried in American political and industrial life. Writing in inexpensive, widely read magazines, these muckrakers, people of the stamp of Ida Tarbell and Lincoln Steffens and Ray Stannard Baker, presented sensational, but thoroughly documented, accounts of the sordid and corrupt doings of big businessmen, labor leaders, and politicians.[32] Upton Sinclair's *The Jungle* was a piece of muckraking that was most significant in encouraging passage of food and drug legislation, and Samuel Hopkins Adams' 1905 exposé of patent medicine frauds was also quite influential. Other reforms, such as restrictions on child labor and a seventeenth amendment to the Constitution, calling for the direct election of senators, were also stimulated by the muckrakers' efforts, but eventually the movement lost its drive. The most exciting stories pall through repetition, and by the outbreak of World War I the public had become inured to tales of social evil. There were other international developments to grip the public interest, and muckraking journalism passed, temporarily, from the American scene. It returned with a vengeance in the 1930s.

The new muckraking was much more consumer-oriented than the original version had been, having taken its origins in a reaction against one of the more distressing trends of the 1920s, the growth of distortion and misrepresentation in advertising.[33] To be sure, the pioneers of advertising, the patent medicine salesmen of the nineteenth century, had played on people's fears of illness to create a demand for worthless, often dangerous, products. In the years following World War One, however, as American business surged forward into its golden age, nearly all industries came to produce more than the public could readily purchase, and more creative advertising became a general imperative. People had to be persuaded, through appeals to such universal ambitions as the desires for youth, health, beauty, sexual success, and social status, that products which in former years might have appeared frivolous were in fact essentials for modern living. From perfumed soaps to more powerful automobiles, a product's quality became a consideration secondary to its public appeal. Competitive advertising created "an uncharted jungle of conflicting claims, skillfully presented misinformation, flattery, sex appeal, and exaggerations"[34] in which the average consumer quickly became lost, to the detriment of his pocketbook and, in some instances, of his health. The clearing of this jungle was the goal of the new muckrakers.

Since some of the most extravagant advertising claims were made for the virtues of non-prescription medicines and cosmetics, since many of these items were hazardous in addition to being inefficacious, and since existing legislation was inadequate to curb these abuses, it was only natural that the focus of the muckraking in the 1930s would be on the frauds and dangers associated with drugs and cosmetics, and with foods as well, and on the need to revise the Food and Drugs Act to remedy the situation. The work that set the standard for the new genre was an almost incredibly successful book with the intriguing title of *100,000,000 Guinea Pigs*. Coauthored by Arthur Kallett and F. J. Schlink, and issued in the first of its more than thirty printings in 1933, the book immediately established itself as the "Uncle Tom's Cabin" of the consumer movement.[35] It was written with nearly as much heat as light, and consequently was guilty at times of exaggerations and distortions suggestive of some of the advertisements for the products it condemned. But it was its sensationalism that recommended the book to so many readers and insured that its message of the dangers in everyday foods, drugs, and cosmetics would be broadcast so widely. Kallett and Schlink established a style imitated by so many other books published soon after that a whole group of authors of the 1930s came to be classed together as "guinea pig muckrakers."

Schlink was a mechanical engineer and physicist who had served for several years on the staff of the National Bureau of Standards. He was drawn into muckraking by his collaboration with Stuart Chase, a certified public accountant turned advertising critic. In 1927 the two published *Your Money's Worth*, an acid-penned guidebook to the "Wonderland" created by "modern salesmanship" in which "the consumer as Alice" was regularly led astray and deceived. The authors' prescription for "Alice" was organization: "The consumer can be organized by the million to jump through the hoops of the advertiser—dosing himself with dangerous nostrums, brushing his teeth with chalk and scent and soap at fabulous prices, walking Heaven alone knows how many aggregate miles for a cigarette. Can he be organized to better his health, increase his real wages, and get a dollar's worth of value for his dollar? Perhaps. . . . It is the consumer's move. If he wants to leave Wonderland, there is a way out, and the clear possibility of drastically reducing the cost of living. He can get his money's worth if he is willing to organize to get it."[36]

Schlink in particular was interested in furthering consumer organization. As an engineer, he appreciated the potential utility of physical science for analyzing the composition, construction, and efficiency of any commercial product. He knew that the consumer, informed with scientific data, could judge for himself the merits of the most misleadingly advertised product. While working on *Your Money's Worth*, Schlink had actually organized a consumer's club to distribute such product information, and his club had met with such an eager reception that in 1929 he was forced to expand and institutionalize his operations. Consumers' Research Incorporated was founded as a nonprofit, unbiased research and educational organization for the examination and testing of consumer products. From the results of a testing program modelled after that of Schlink's former employer, the National Bureau of Standards, Consumers' Research rated products as "recommended," "intermediate" or "not recommended," and passed these ratings on to subscribers of its bulletin.[37] In the first four years of the organization's existence, the number of subscribers mushroomed from 1,000 to 45,000.[38]

Such rapid growth would indicate that *Consumers' Research Bulletin* offered something more than a bland listing of analytical data and recommendations. Schlink, in fact, regarded his *Bulletin* as above all a pulpit from which to preach the new consumerism, from which to warn the people of the snares and pitfalls laid by big business and to castigate transgressors against the consumer's right to "a dollar's worth of value for his dollar." Sins of omission were judged as harshly as those of commission. His book with Chase had bemoaned the weakness of federal agencies responsible for consumer protection, and Schlink was especially dismayed by the placement of the Food and Drug Administration within the Department of Agriculture. Conflicting interests would keep the FDA forever enfeebled, he argued, unless it could be reformed into an independent Department of the Consumer: "The Food and Drug Administration always feels obliged to apologize to business interests for any slight aid it may give the consumer. It is high time that the functions of this bureau were transferred to the Department of the Consumer proposed by CR so that these functions may be vigorously and unashamedly performed for the benefit of taxpayer-consumers as a whole."[39]

There was no better example of the FDA's mixed obligations than spray-residue regulation, and it was through the medium of *Consumer's Research Bulletin* that many citizens were first alerted to the dangers of arsenic and lead on produce.[b] In a 1932 article on the hazards of metallic poisoning from foods, Schlink mentioned that "most people are probably unaware of the risks of arsenic and lead poisoning they incur in eating fresh fruit."[40] Subsequent issues of the *Bulletin* dealt with residues much more fully and seriously,[41] but it was in *100,000,000 Guinea Pigs* that the problem was given its most significant national airing.

The book's second author, Arthur Kallett, was also an engineer, and a board member of Consumers' Research.[42] He, like Schlink, had become horrified by the number of

[b] An earlier public discussion of the residue hazard had been presented in *Hygeia*, the popular health magazine published by the American Medical Association, in 1927, but it had not been an article to produce much excitement. The possibilities of chronic arsenicism and plumbism were noted, but the necessity of spraying for production of sound fruit was given at least equal emphasis, and the author concluded with the assurance that "all competent authorities agree that there is nothing in the situation to be alarmed about." E. S. Clowes, *Hygeia*, 5, 461 (1927).

189

harmful foods, drugs, and cosmetics that continued to flood the market in spite of the Food and Drugs Act. As will be discussed later, most of the fault for this situation lay with the law, which was riddled with loopholes. Kallett and Schlink were aware of the act's inadequacies, and began their book with a delineation of them, but they were also convinced that the FDA was failing fully to employ the legal tools it did have because of an undue regard for the financial interests of the industries it was supposed to regulate. As the result of this weak enforcement of a weak law, American citizens, more than one hundred million of them, were "all guinea pigs, and any scoundrel who takes it into his head to enter the drug or food business can experiment on us."[43]

In the pages that followed, the authors exposed the scandalous results of dozens of these "experiments." The bleaching of fruit with sulfur dioxide and of flour with nitrites, and the preservation of foods with benzoates and sulfites, were charged with lopping five to ten years off the average American's life expectancy; highly touted antiseptics, such as Listerine, were reported to be "practically worthless as protection against infection";[44] cosmetic products like Koremlu, a thallium-containing depilatory, and Kolor-bak, a hair-dye made from lead acetate, were scored for subjecting unsuspecting women to dangerous poisons. As early as the book's third chapter, under the heading, "A Steady Diet of Arsenic and Lead," Kallett and Schlink introduced their readers to the hazards of spray residues. It was one of their shorter chapters, but it was punctuated with as high a concentration of exclamation marks as any, and was thoroughly chilling. "Six persons poisoned in California in 1931 by greens sprayed with lead arsenate," it began. "A four-year-old Philadelphia girl dead, in August, 1932, from eating sprayed fruit. With a background of cases like these, are you willing to have even very small doses of arsenic, a deadly poison, administered to you and your children daily, perhaps several times daily? Willing or not, if you eat apples, pears, cherries, and berries, celery, and other fruit and vegetables, you are also eating arsenic, and there is good reason to believe that it may be doing you serious, perhaps irreparable, injury."[45]

The origin of this arsenic in insecticide sprays was then explained, the gradual development of residue regulation outlined, and the authors' contention that the FDA tolerance was still too lenient supported by quotations from medical authorities. Finally, lead poisoning from residues was presented (again with the aid of medical testimony) as an even more serious threat than arsenic poisoning. Kallett and Schlink left no doubt that spray residues were a major menace to the public's health, and that they were allowed to remain so only through the negligence of the FDA: "Nothing but ordinary official complacency prevents the responsible officials from seeking a remedy for the spray residue situation. Even though the problem concerns only protection for public health and not increased income for farmers and fruit growers, it should nevertheless be possible to find funds and experts to reach a solution."[46]

The only way to force a solution to the residue, and to other food, drug, and cosmetic hazards, they concluded two hundred pages later, was for each and every American "guinea pig" to stand up and demand of his local and national legislators and health officials that a much stronger consumer protection law be enacted, and that it be strictly enforced: "Above all, let your voice be heard loudly and often, in protest against the indifference, ignorance, and avarice responsible for the uncontrolled adulteration and misrepresentation of foods, drugs, and cosmetics. In this adulteration and misrepresentation lurks a menace to your health that ought no longer be tolerated."[47]

The response to this call to arms was rather gratifying, considering all the other issues competing for the attention of a public mired in a severe economic depression. Certainly the volume of mail to the FDA critical of the handling of the residue problem increased sharply after the publica-

tion of *100,000,000 Guinea Pigs.* Kallett and Schlink's inclination toward alarmism clearly rubbed off on many of
their readers. An excitable gentleman from Salt Lake City,
for example, wrote: "in the interest of the children of
America and their foods. My heart goes out to them. For
many years they have never had natural foods to eat. Natural foods raised under natural law, Gods law. . . . No arsenical spray of contamination. Poison in every form should be
prohibited. . . . Miller Cahoon lost twenty head of fine
horses and mules by feeding hay that grew under trees that
had been sprayed with arsenic spray. These and many more
died with heart failure.

"And here is my complaint. The death rate in America
just out, shows thirty-eight thousand more people died in
1933 than in 1932, and a great majority died with heart
failure. We are wondering if these people died under the
same conditions as the horses. Are the college professors
boosting sales of arsenic spray to help the smelters sell their
arsenic? Now arsenic is a slow but deadly poison and should
never be allowed in any sprayed on foods. Pure foods can
and if rightly raised under natural law, Gods law. To poison the dawning of a new day when the insects are taken
away the poison.[?] . . . Pure foods makes pure people. More
than thirty-eight thousand people dying unnatural deaths
should be looked into. . . . Yours for a better America and
pure foods. . . ."[48]

Recognizing the need for haste in purifying the food supply, a Miami woman appealed directly to the nation's first
lady:

"My dear Mrs. Roosevelt:

"Asking your kind indulgence, I want to tell you that I
thank God for having placed your noble husband at the
head of our government. . . . I realize the great problems
he has before him, and therefore hesitate to write to him
regarding the one which I am about to present to you, although I cannot but believe that others must have brought
the same subject to his attention.

"While chemists the world over are feverishly working to concoct poisonous gases for the wholesale destruction of their fellow-men—which all right minded people condemn, our farmers and horticulturalists are poisoning our own people with deadly insecticides. Compounds of arsenic are in common use, and one of these—lead arsenite [sic]—is sold in enormous quantities, estimated to be about one-half pound per capita per annum. . . . It seems to me the chemists of the country should be set to work to discover an effective insecticide that would be harmless to humanity. And laws should be made putting a stop to the use of these poisonous applications. . . . And surely something should be done to stop the use of these poisons on our fruits and vegetables, from the effects of which many people have lost their lives.

"With deep appreciation of all that you have done for our people, and hoping something can be done in this matter; and wishing you and our noble president a very Happy New Year, I remain. . . ."[49]

The new level of consumer unrest created by Kallett and Schlink's book was to be sustained by one sequel after another, quickly contributed by other guinea-pig muckrakers. Schlink himself wrote, *Eat, Drink and Be Wary* in 1935, a book less concerned with berating the Food and Drug Administration and effecting improved legislation than it was with liberating people from the detrimental artificialities of modern diet. It was Schlink's conviction that everyone would be better off if he would "follow your grandmother's instincts" and go "back to the . . . pre-Crisco days," repudiating "artificial, preserved, stabilized, canned, ethylene-ripened, dyed, and over-processed foods too far removed, in growing and marketing, from sun and air and the realities of woodland, farm, and field."[50] His book was also largely a defense of the flesh diet against the criticisms of vegetarians, but tucked away between his thrusts at bleached flour and at raw vegetables was a chapter entitled, "Some Very New Means to Poison Us All," which described how "the fruit and vegetable experts prefer arsenic to worm-holes." As he had done with Kallett, Schlink outlined the farmer's use of arsenical sprays, indicated that residues from the spray were commonly present on produce, and then drew freely from the medical literature to support his charge that the residues were dangerous. To clinch his case, Schlink appended a list of nearly two dozen different fruits and vegetables and the arsenic and lead content for each, followed by a recitation of acute and fatal poisonings caused by items ranging from sprayed apples to sprayed broccoli to sprayed celery.

As frightening as was the Schlink book, a 1937 publication by Rachel Palmer and Isadore Alpher was even more disturbing. *40,000,000 Guinea Pig Children* probed the most irritable area of public sentiment, its solicitude for innocent children. The authors deplored the invention of a special children's market by the insatiable advertising industry, and cautioned parents about the many hidden dangers in children's products. Since children have a particular fondness for fresh fruit, the spray residue hazard was also given detailed consideration. And if the advice that

> *Any bug that eats lead*
> *Will soon need a casket;*
> *Children should skin*
> *The fruit in the basket.*[51]

was doggerel, it nevertheless effectively carried the point that insecticide residues could dispatch children as readily as bugs.

FDA personnel read these lines, and the messages of the rest of the guinea-pig writers, with mixed feelings. They were angered by what they considered the unfair evaluations of their performance in food and drug control. Kallett and Schlink had asserted that, with regard to spray residues, "The attitude of the Food and Drug Administration has been little short of brutal indifference."[52] In reality, FDA officials believed it had not been their indifference, but rather technological and political barriers, that had retarded solution of the residue problem. Farmers had claimed to be unable to meet the desired tolerance restrictions immediately, and their congressional representatives had threatened to remove, even outlaw, such restrictions altogether. A very gradual implementation of the proper tolerances had thus seemed the only feasible policy.

But, the guinea pig authors demanded, need this policy have been conducted in such secrecy? Did not the FDA have an obligation at least to warn the public that most produce carried potentially injurious quantities of arsenic and lead so that people could attempt to provide themselves with the protection that the FDA was failing to give? These were questions the FDA had feared for years. It had dreaded the day it might have to answer for the agreement with the agricultural industry not to publicize the residue hazard. The concession had been one made in good faith, in the belief that this consideration would encourage agriculturalists to put their house in order more quickly. But they had not done so, and when the muckraking began the FDA was left standing alone, embarrassingly stripped of any solid defense. "Government officials have . . . suppressed important information on the arsenic hazard," Kallett and Schlink announced, "and have resisted in every way the opening up of the question to discussion in the interest of public safety. The Government has acknowledged the hazards of excessive consumption of arsenic residues; it has permitted residues large enough to constitute a serious health hazard; yet we cannot find that it has uttered one word of warning to the public, or even so much as suggested mildly that apples and pears be peeled before they are eaten. . . . What business loss, Mr. Campbell, is equivalent in the Adminis-

195

tration's arithmetic to the poisoning of hundreds of thousands of citizens, from babyhood to old age, with arsenic-spray residue?"[53]

If the guinea-pig books angered and at times humiliated the FDA, however, they also were welcomed as potent new weapons for the battle for improved food and drug regulation. Campbell's chief assistant suggested to him that although "Schlink had no intention of posing as an ally of the Food and Drug Administration . . . unconsciously he may be a very valuable one." The value of *100,000,000 Guinea Pigs,* another observer appreciated, lay "in its calling attention to the rottenness and inadequacy of our laws."[54]

The 1906 Food and Drugs Act was, in fact, if not rotten, at least highly inadequate, and FDA officials, as much as the guinea pig journalists, had worried about these inadequacies for some time.[55] The burgeoning field of cosmetic preparations, for instance, was completely unaffected by the act. In 1906, the cosmetic industry had been too small in relation to the food and drug industries to excite much concern. By the 1930s products purporting to relieve every conceivable defect of attractiveness were being peddled far and wide, and many of them were distinctly hazardous. Yet when a young woman was permanently blinded by a commonly used eyelash dye, or another's face turned a dirty blue color by her freckle remover,[56] neither the FDA nor any other governmental agency could institute legal proceedings against the products' manufacturers.

The ever-popular patent medicines were regulated by the Food and Drugs Act, but not fully. The law, first of all, did not prohibit the inclusion of toxic ingredients in these non-prescription remedies. Knowing that the American citizen, proud of his independence, would demand the right to select his medications for himself, the framers of the law had sought only to inform him of any risks associated with a remedy by requiring that certain poisonous drug sub-

196

stances be listed on the label.^c Nor could drug manufacturers claim on their labels that any ingredient was present when it was not. Thus provided with reliable information, it was assumed, the consumer's common sense would direct him toward safe purchases. The mislabelling provisions of the act nevertheless failed on several counts. Dangerous drugs other than the substances originally required for labelling, such as the barbiturates, came to be incorporated into patent medicines. In 1911, the U.S. Supreme Court ruled that the act's prohibition of any label statement that was "false or misleading in any particular" was not applicable to therapeutic claims, that as long as the manufacturer properly listed any legally specified ingredients in his remedy, he could claim it as effective treatment for any or all ailments. President Taft himself (not to mention Harvey Wiley, of course) was so outraged by this interpretation of the act that he personally prodded Congress to mend the loophole. The resulting amendment, the work of Kentucky's Congressman Sherley, misfired. It prohibited any remedy's package or label from carrying "any statement, design, or device regarding the curative or therapeutic effect of such article or any of the ingredients or substances contained therein, which is false and fraudulent."[57] Demonstrating that a product's claim to cure everything from warts to cancer was false would usually be a simple matter, but proving that, in addition, the manufacturer knew the claim to be false, that the claim was intentionally fraudulent, could be virtually impossible. Hence, even armed with the Sherley Amendment, the FDA's hold on patent medicine labelling was weak, and it had no control whatever over false advertising through media other than the labelling.

A final inadequacy of the act was its ambiguity of lan-

c These were alcohol, morphine, opium, cocaine, heroin, alpha and beta eucaine, chloroform, cannabis indica, chloral hydrate, and acetanilid, and any derivatives or preparations of these substances.

197

guage with regard to food adulterations (noted previously in Chapter 4). The FDA still lacked the power to set legal standards for food composition or excessive levels of contamination.[d] This particular weakness was the one that most seriously hampered spray residue regulation, for without the authority to establish arsenic and lead tolerances having the force of law, every contested seizure of sprayed produce had to be defended in court, and the defense was not always successful.

It was these flaws in the Food and Drugs Act that allowed the development of a situation by the 1930s in which the American consumer was in fact a guinea pig. But Kallett and Schlink and their imitators were not alone in condemning the act. Its failings had been cause for complaint by any number of officials within the Bureau of Chemistry, and then the FDA, and by no one more than Walter Campbell. In 1933, shortly after reading *100,000,000 Guinea Pigs*, Campbell wrote a colleague that "I have been pursuing for years a policy calculated to correct these defects by seeking amendatory legislation. If this muckraking publication furthers these ends, it will not have been published in vain."[58] This and other muckraking publications were significant in generating a demand for a better law, though the level of popular disturbance due to the guinea-pig literature must not be overestimated. Only a minority of the public became actively involved with food and drug reform, while most people overlooked the problem in their concern to meet the more immediate troubles of the Depression.[59]

The decision actively to seek a new food and drug law was actually crystallized by events within the government, events in which the spray-residue question was a precipitating factor. Early in 1933, the Department of Agriculture received one of those letters that were becoming increas-

[d] The 1930 McNary-Mapes Amendment to the Food and Drugs Act gave the Secretary of Agriculture the power of setting minimum standard of quality and fill for *canned* goods.

ingly frequent, a consumer's inquiry about what the government was doing to protect the public from spray residues.[60] The letter was referred to the FDA for reply, and the reply returned to the Secretary of the Department for his signature. The procedure had already become routine, and the FDA's response was much like the many others it had written: "The Department shares your concern as to the possible effects on public health of the practice of spraying agricultural food crops with certain poisonous insecticides. Let me say, however, that the Department is devoting a very material proportion of its available funds and personnel to this problem in the interest of public health and the welfare of the fruit and vegetable producing industry."[61]

The routing of this particular letter from the Secretary's office to the FDA and the return of the reply, however, happened to occur during the changeover from Hoover's administration to Roosevelt's, and by the time the standard FDA reply reached the Department's office, a new Assistant Secretary of Agriculture had been installed, one who found the reply unsatisfactory. Assistant Secretary Rexford Tugwell was an enthusiastic young "brain-truster," a man eager to use his learning as an economics professor for the improvement of American society. He was not a man inclined to compromise, and he wasted no time returning the FDA letter unsigned but with an attached memorandum: "This seems to me an unnecessarily elaborate argument if our position is sound on this; I am not satisfied that any part of our argument should rest on protection to a producing industry. The law, I believe, does not contemplate this administrative discretion."[62]

The Tugwell note struck Campbell, who had struggled for years with the residue problem and its attendant political pressures, as "a kick in the teeth."[63] He was accustomed to the charge that his FDA was overly considerate of economic interests, but it was one thing when the charge was levelled by a muckraker, and quite another when it came from one of his superiors in the Department of Agriculture. It was largely because of the Department's commitment to aid the agricultural industry, after all, that the FDA had been forced to extend such leniency to fruit growers in the first place. Obviously the new Assistant Secretary was ignorant of the limitations of both the Food and Drugs Act and of the FDA's administrative situation, and Campbell immediately crossed the street to see Tugwell and set him straight.

Their meeting proved considerably more peaceful than Campbell anticipated. Tugwell became most sympathetic once the facts were placed before him and he ended by agreeing with the FDA chief that the 1906 law needed revision. By the time Campbell ate lunch and returned to his own office, Tugwell was summoning him back to announce that since their morning interview he had talked with the President and received approval for a revision of the Food and Drugs Act. The chain of events connecting this decision to revise the old law to the subsequent resolution to draw up an entirely new replacement for it, to the enactment, more than five years later, of the Food, Drug and Cosmetic Act, need not be recited here. The story has been covered fully and soundly in several books,[64] and involves a number of issues not directly pertinent to the development of spray-residue regulation. The immediate import of the realization that the time had come for improved consumer protection lay in the educational measures adopted by the FDA to rally support for a new law. The guinea-pig literature had not yet reached full flower, and most people still assumed that because there was a Food and Drugs Act on the statute books, the safety of all food and drug products was guaranteed. This innocence would have to be destroyed, and a public clamor for better protection created, if the opposition to the law certain to come from food, drug, and cosmetic producers was to be overcome. Thus FDA personnel had to sandwich the long hours of work necessary to draft a foolproof food and drug bill between the hours spent delivering radio talks and writing pamphlets explaining to the public why a new pure food law was needed. In conducting this educational campaign, the FDA broke with its tradition of sheltering the residue problem in secrecy. Indeed, thanks to the activity of muckrakers, the secret was already out, and the "steady diet of arsenic and lead" so shrilly denounced by Kallett and Schlink was soon being publicized, if more quietly, by the FDA as well.

The reaction of the agricultural industry to this publicity is curious. After the flogging that farmers had taken from guinea-pig sensationalism, they might have been expected hardly to notice, or perhaps even welcome, the restrained discussion given spray residues by the FDA. Instead, they were made furious by what seemed the FDA's betrayal of trust. Two instances of breach of promise were especially outrageous. The first was a radio broadcast of February 18, 1935, an installment of "Uncle Sam at Your Service," a weekly report from the Food and Drug Administration.[65] The ten-minute program began with the announcer's observation that:

"It seems to me, after reading today's message from the Food and Drug Administration, that something ought to be done about those verses we learned in school in kindergarten days.

"The first page of my primer showed a beautiful apple tree, with fruit perfect in shape and color, free from the slightest blemish. Under the beautiful apple tree, in big type, was this verse:

'A is for Apple,
So round and so red,
That grew on the apple tree
Over your head.'

Only that, and nothing more. It never occurred to me that somebody besides Mother Nature was responsible for those apples. I didn't know the owner of the apple tree had to fight insects and blights and rots—and many other apple troubles. I didn't know he had to spray those trees with a poisonous chemical, so we'd have enough apples to fill our lunch boxes and our pockets."

These gaps in his knowledge, the announcer went on to imply, had only just been filled by the FDA.

"I'm not a versifier, but after reading this report, which deals with spray residues on apples and pears, I'm recommending a change in apple verses for modern youngsters. The 'A' page of a primer should include something like this:

'*A is for Arsenate—*
Lead, *if you please,*
Protector of apples
From arch enemies.'

Yes, sir—and madam—you might as well bring up the younger generation to know that it takes a lot of work to provide them with Jonathans and Rome Beauties and Winesaps—as well as Bartlett pears, and Seckle pears, and Keiffers."

To this point, the program had offered nothing but praise for the intelligence and industry of the nation's fruit growers. It now took an unsettling turn. Lead arsenate, the broadcaster continued, not only killed insects but could, if left on produce sent to market, cause chronic illness in people. "Fortunately for us, the growers can keep down to a safe limit the amount of the poisons occurring on their fruits and vegetables," he went on, and should one occasionally fail to remove his residues adequately, the constant vigil maintained by the FDA's inspectors would keep the food from ever getting to market. This theme was then repeated until the announcer brought the show to an end with the cheering assurance that "we need not worry about spray residues on apples and pears, as long as inspectors of the Food and Drug Administration are on the job."

It had been a seemingly innocuous program, a bit too

condescending perhaps, but fully in keeping with the FDA's goal of producing a balanced picture of the food and drug situation, of offsetting the distortions of the guinea-pig literature while acknowledging the hazards that did exist. In the broadcast, the danger of residues had been minimized, and the necessity of spraying and the integrity of growers defended. The irony of the show, in fact, was that it had been intended as a peace offering to agriculturalists. "The purpose of this talk," a baffled Assistant Chief Paul Dunbar recalled several weeks later, "was to reassure consumers who had been frightened by exaggerated stories of danger to their fruit supply and to point out the necessity for the use of poisonous insecticides in growing apples and pears. Much to our surprise we received a number of complaints from members of the fruit industry."[66]

The number of complaints was considerable, and did not come entirely from fruit growers. Even some employees of other bureaus of the Department of Agriculture were annoyed by the interference created for their work by the broadcast. A field representative assigned to advise Yakima district orchardists on codling moth control, complained: "This publicity (considered by growers to be unnecessary), together with the way the subject was handled, has caused serious irritation among the growers of the Northwest. Broadcasts of this kind, which cause the growers to become antagonistic to the Department, are very embarrassing to our men in the contacts with the industry. The goodwill of the growers toward the Department is an important factor in the success of our work. It is not for us to express an opinion on the necessity of such a broadcast from the standpoint of the Department's obligation to the consumer. We urge very strongly, however, that when radio talks of this kind are under consideration, full weight be given to the probable reaction of growers to them and its resultant effect on the relations existing between Department representatives and the industry."[67]

The effect of the 1935 radio broadcast on relations be-

tween the Department and the fruit industry was nevertheless mild compared to the indignation excited by a book
published the following year. In 1933, when the work of
drafting a new pure food law and getting it through Congress was begun, FDA personnel started a collection of
samples of dangerous drugs and cosmetics not covered by
the old law. Intended to be used for purposes of illustration at Senate hearings on food and drug legislation, the
items were in the meantime stacked around the walls of
Walter Campbell's office. There they caught the attention
of newspaper reporters visiting FDA headquarters, and the
reporters were so startled by the array of familiar bottles
displayed as death dealers that they began referring to the
collection as the "Chamber of Horrors." The propaganda
potential of the "Chamber" was manifest, and before long
an exhibit of many of the more dangerous and fraudulent
products, mounted on beaver board and accompanied by
printed descriptions of their horrors, was being circulated
about the country and impressing crowds everywhere with
the inadequacies of the Food and Drugs Act.

The exhibit impressed Ruth deForest Lamb, the FDA's
information officer, with the desirability of a book docu-
menting the same horrors and inadequacies. Her *American
Chamber of Horrors*, issued in 1936, was not an official
FDA publication, but it made extensive use of the Administration's files to compile a shocking indictment of the
country's food, drug, and cosmetic industries. Miss Lamb
was not just another guinea-pig muckraker, for, while her
book revealed some alarming abuses, it did not share the
Kallett, Schlink, et al. opinion that the continuation of
these abuses could be blamed on an unconscionable FDA.
American Chamber of Horrors was at once an exposé of
industrial immorality and of governmental powerlessness.
The tragedies the book recounted, "and countless others,
have actually happened. They have happened, not because
Government officials are incompetent or callous, but because they have no real power to prevent them."[68] The

Lamb attack was not on the FDA, but against the weak law with which it had been armed.

The longest chapter of *American Chamber of Horrors* was the one entitled, "How Much Poison is Poisonous?" In its fifty-five pages, the author repeated the now familiar story of spray-residue regulation since 1919, but with a new interpretation. What had seemed extreme leniency to Kallett and Schlink had in fact been a policy of compromise forced upon the FDA by political and economic factors associated with the residue situation. Far from being lackadaisical bureaucrats shirking their responsibilities to the public, Lamb depicted her FDA colleagues as dedicated men who had made the very best of a bad bargain. Her accounts of the confrontations of Western District inspectors with fruit growers angered to the point of violence even raised some Administration personnel to heroic stature.

But the spray-residue chapter evoked as much fright as it did admiration. It told of "the little Scribner boy," who was ill for six months after eating too many sprayed pears; of "ten-year-old Ralph Dodge," who was killed by sprayed apples; of "little Naomi Wilkin," a victim of a neighbor's sprayed grapes. "But the real risk from arsenic and lead in spray residues is not acute poisoning! Rather, the danger lies in the slow, insidious undermining of health from the accumulation of the metals in the soft tissues and bones."[69] By the time Miss Lamb completed her catalogue of residue horrors, none but the most hardened could even look at a piece of fruit without apprehension.

American Chamber of Horrors was a best-seller; 20th-Century-Fox even considered converting it to a screenplay.[70] The producers of foods, drugs, and cosmetics considered rather different treatments of the book, and, though the wholesale burning of all copies was impossible, the FDA could at least be upbraided for having cooperated with its author. The vituperation heaped on the Administration flowed from all quarters of the insulted industries, and fruit growers certainly contributed their share. The general feeling in orchard districts was that *American Chamber of Horrors* was an escalation of the FDA's (or perhaps even the Department of Agriculture's) campaign to destroy the western fruit industry. A Washington grower actually sent her denunciation of the "gang of leeches . . . from the Pure Food Dept. at Washington, D.C. . . . living off this residue racket" to Eleanor Roosevelt, with the request that she personally deliver the letter to the President. But by all means, "Please don't make the mistake of sending this letter to the Department of Agriculture."[71]

A tendency toward paranoia was observed in the examination of the agriculturalists' attitudes toward residue regulation during the late 1920s. As the 1930s advanced, these signs became increasingly pronounced. Suspicions of persecution or conspiracy could be planted in the growers' minds by the mildest disturbance; it did not take a Schlink or a Lamb to send a farmer running from his orchard to his writing desk to appeal to his congressman or another government official for relief from unwarrantable oppression.

An inclination to see themselves as objects of governmental discrimination and plotting is evident in the agriculturalists' reactions to the 1935 radio discussion of residues, and in other teapot tempests of their own making. Early in 1933, for instance, Department of Agriculture chemists at last succeeded in developing a rapid and accurate method of determining quantities of lead in spray residues on produce.[72] Up to that time, lead had been regulated with no more than a wish, that the arsenic tolerance was too low to allow significant amounts of lead to remain from the lead arsenate spray. With the new analytical method, it became possible to regulate lead residues directly, and on February 21 an administrative tolerance for lead of 0.025 grain per pound was announced to the agricultural industry. The subsequent history of the lead tolerance will be discussed in the next chapter; for the present, the immediate reaction of growers to the announcement is important. In spite of the fact that the dangers of lead had been stressed at spray-residue conferences for years, and that the FDA had promised long before to institute lead regulations as soon as a satisfactory analytical technique could be perfected, fruit growers, especially those of Washington state, interpreted the lead tolerance announcement as the latest scheme devised by the "residue racketeers." "When they found that we could clean the arsenic," a Wenatchee orchardist surmised, "the order came to forget the arsenic and clean the lead."[73]

The furor that arose around the discovery of arsenical residues on tobacco offers another example of the fruit growers' suspicions of the government's motives. Another of the crops to which Paris green began to be applied during the late nineteenth century was tobacco and, as usual, the application was often careless. By the mid-1890s reports had been published of considerable arsenic residue on tobacco leaves and warnings to farmers not to overspray the plant.[74] The early reports drew little attention; tobacco came to be treated with newer arsenical insecticides; and not until 1927 was another warning of "a hitherto unsuspected source of arsenic in the human environment"— tobacco—issued.[75] Studies revealing that American tobacco carried much higher levels of the poison than did the tobacco of other countries suggested lead arsenate insecticide as the source of the arsenic,[76] and both experimental and clinical investigations indicated that arsenical cigars and cigarettes might be hazardous to health.[77]

Concern over arsenic in tobacco received sufficient coverage in the agricultural press to alert fruit growers to the problem and then to lead them to ask why, if their fruit was being confiscated as dangerous, tobacco was permitted to pass unmolested through interstate commerce. Washington growers refused to let this prejudicial enforcement go uncontested, and at the 1937 meeting of the Washington State Horticultural Association they succeeded in getting adopted a resolution for presentation to the FDA:

"WHEREAS the apple industry has had exceedingly severe restrictions with reference to the presence of arsenic by the Federal Food Department [sic], and

"WHEREAS tobacco, whether pipe or chewing tobacco, usually carries arsenic in amounts many times that allowed on apples, and

"WHEREAS tobacco is much more widely and continuously used than apples, and

"WHEREAS in smoking tobaccos much arsenic is taken into the system through the respiratory tract, a recognized far more serious ingress.

"THEREFORE BE IT RESOLVED that we demand that the government desist from such discrimination and refrain from capriciously enforcing this regulation in regard to apples unless the law can be equitably inforced [sic] in regard to all products."[78]

Walter Campbell's reply that "none of the provisions of this statute [the Food and Drugs Act] relate to tobacco or tobacco products" and that "the Food and Drug Administration has absolutely no jurisdiction over tobacco"[79] hardly affected the flow of letters from fruit men demanding to be given fair treatment.

The aftermath of a seemingly harmless 1936 play is perhaps the most telling indication of the orchardists' growing testiness. *Russet Mantle* was the work of Lynn Riggs, a playwright whose only enduring fame derived from the inspiration his earlier play, *Green Grow the Lilacs*, provided for the writing of *Oklahoma*. To judge from the reviews, the humor of *Russet Mantle* was considered engaging at the time, but its attempts at dramatic social commentary seemed as banal then as now. Its comedic appeal nevertheless carried the play through 117 performances on Broadway, after which it ran for several lean weeks in San Francisco and Los Angeles, then quietly exited for good.

It took time for the news of Broadway to trickle down to the farm, but once word of *Russet Mantle* reached fruit-producing regions, the response to its message was spontaneous and considerably more animated than that any big city audience had given it. The play seemed an outrage. It was not its stale story of youth in revolt, of a radical young poet who recruits a free-spirited farmer's daughter to his mission of creating a new world, that offended fruit growers. Rather, they objected to the play's one saving grace, its comedy, and in particular to the author's intended humorous lines at the very opening of Act I. Horace Kincaid, Sante Fe rancher and a symbol of the older, misguided generation, appears on stage first, sampling some apples brought him by his farmhand Pablo.[80]

> *Horace* (absurdly, like a patient professor): You see, Pablo? Every year—better apples, better plums, better peaches. That comes of spraying. The blight and the bugs can't stand the spray. The fruit has a chance to develop. Do you understand?
>
> *Pablo*: The spray, yes.
>
> *Horace* (as before): The spray does it. The blight vanishes. The bugs die.
>
> *Pablo* (with pleasant sorrow): Si. The leetle bugs—they doan like to eat poison. It makes them seeck. They crawl—on three legs, on two legs, they curl up in their holes, they croak dead.
>
> *Horace*: Exactly!
>
> *Pablo*: Poor leetle bugs.
>
> *Horace* (faintly annoyed): I wouldn't waste sympathy on them, Pablo. It's apples or bugs. Take your choice.
>
> *Pablo* (unhesitatingly): Oh, I take apples!
>
> *Horace* (a minor triumph accomplished): Of course. And good apples, too.

After several more exchanges, Horace takes a bite of an apple, makes a face, spits the apple out, and asks Pablo what had given the fruit that horrible taste.

209

Pablo: The spray, I theenk.
Horace (gulping): You've poisoned me. Ugh!

Horace quickly rinses his mouth with water, at which point his wife Susanna comes in complaining that her chickens are not laying enough eggs.

Susanna (as casual as she can make it): There's one comfort: I'm better at chickens than you are at fruit.
Horace (outraged): Better!
Susanna (sweetly venomous): At least you can take a bite of one of my broilers without getting poisoned.
Horace: Susanna! That's not fair.
Susanna: It's true, though.

Susanna's sister Effie, a Southerner characterized by "mindless cheerfulness," now enters the conversation:

Susanna: His fruit trees don't make a cent, Effie—not a cent!
Horace: But they will. You wait and see! The world will hear of my orchard yet!
Susanna: Umh! I can just see the headlines. "KINCAID APPLES POISON SEVEN HUNDRED."

Several minutes later, the hero, a young wanderer named John, shows up in search of work. To gain Horace's favor, John tells him that he has sampled some of his apples while coming down the road, and that they are better than any he has ever had, even those of the Northwest.

Horace (cordially): I'm glad to hear you say that, young man. (Remembering in alarm) Oh! I hope you wiped the spray off first. It's poison, you know.
John (smiling): Oh Lord, yes! I'm an old hand at stealing apples.
Horace (politely): Where have you done most of your apple stealing—in the Northwest?

The characters then moved on to discuss weightier matters, and spray residues were not so much as mentioned again. But the orchardists reacted as if the whole play, which took its title from Hamlet's description of the morn "in russet mantle clad," were about the mantle of arsenic covering their russet apples. Complaints that the play had "depicted that sprayed apples had killed 700 persons" rained down on the FDA. Even men so highly placed as the Director of the Washington [state] Department of Agriculture denounced *Russet Mantle*'s "ill-advised dramatization of the spray residue question."[81]

Throughout it all, from the farmers' petitions to regulate tobacco to their demands that *Russet Mantle* be closed down, the FDA was also receiving pleas from frightened consumers urging that stricter residue controls be imposed. FDA officials were caught in a crossfire, but action was already being taken along several fronts to get out of it.

The conferees consider it to be a matter of
fundamental economic as well as social and
health importance to the food industry of this
country . . . that researches be pushed vigorously
through the resources of the Government
in order to discover a substitute for lead arsenate as
an insecticide and fungicide for fruits and
vegetables. The conferees suggest the desirability
of the Department's undertaking, either
itself or through other branches of the Federal
Government, experiments upon the chronic
poisonous effects of lead arsenate ingested
with or without raw and cooked fruit and pulp
juices, and that studies be made of the desirable
effective maximum and minimum amount of
lead arsenate as applied to and recovered
from fruits and vegetables.

—Report of the Hunt Committee, 1927

"No Longer a Hazard" 7

THE BEST POSSIBLE SOLUTION to the spray-residue problem
had always been obvious. Even before lead arsenate was
introduced in the 1890s, entomologists had hoped that
organic pesticides, such as pyrethrum, might be perfected
as replacements for Paris green and London purple. The
expectations for pyrethrum proved overly sanguine, but
the desirability of an insecticide as effective and as inexpensive as arsenic, without that poison's toxicity to higher animals, remained clear. As uneasiness about arsenical residues resurfaced during the 1920s, the demand for new

insecticides also assumed a new intensity. In the same year that the Hunt Committee advised the government to direct researches into substitute insecticides, the editors of *Industrial and Engineering Chemistry* observed that "entomologists have not regarded as serious [sic] as it deserves the insidious character of lead and arsenic," and proceeded to "sound the alarm" to summon chemists to the urgent search for insecticides less harmful to man.[1]

The events of 1927 also stirred up among farmers interest in new insecticides. Fruit growers who had been happily spraying lead arsenate for years began to appreciate, soon after the announcement of the Salt Lake City Spray Conference proceedings, that an effective non-arsenical could relieve them of the annoyances sure to accompany enforcement of the new arsenic tolerance policy. As the experiences of the next few years revealed just how frustrating tolerance restrictions could be, agricultural insistence mounted for a replacement for lead arsenate. The Tennessee entomologist Simon Marcovitch acknowledged in 1930 that the demand for an arsenical substitute had come largely from "apple growers who are harassed with the residue problem."[2] In the same article, Marcovitch announced that seven years of searching had uncovered two fluorine compounds, cryolite (Na_3AlF_6) and barium fluosilicate ($BaSiF_6$), which promised to meet the specifications for a lead arsenate substitute.

One of these specifications, of course, was a very low chronic toxicity for man. The prospects here seemed bright, since a 1921 study of fluoride intoxication in rats, conducted by the prominent toxicologist Torald Sollmann, had indicated that the chronic effects of appetite and growth retardation were produced only by amounts of fluoride considerably greater than those to be expected as insecticide spray residues.[3] The chronic toxicities of cryolite and barium fluosilicate were nevertheless examined soon after Marcovitch's introduction of the fluorine compounds. The H. F. Smyths, a father and son toxicologist team, compared the chronic effects of the new insecticides with those of lead arsenate, and found the arsenical insecticide to be about eighteen times more toxic for rats than either of the fluorine compounds.[a] The Smyths optimistically concluded: "It would seem that the use of fluorine insecticides would leave a much wider margin of safety than do arsenical materials between the weight of spray residue on fruit and the amount toxic to the consumer."[4]

Insecticide manufacturers did not miss their cue. Both of the fluorine insecticides were quickly brought to market and were being used in the field by a number of the more progressive western orchardists by the 1933 season. FDA officials, however, were dismayed by this development, and particularly by "the impression given in some quarters that fruit and vegetable growers could turn away from lead arsenate toward the fluorine compounds with the assurance that their troubles with regulatory officials would be over."[5] What the FDA knew, and farmers were soon to learn, was that a switch to cryolite might be only a leap from frying pan to fire. The early 1930s were also the years in which the condition of mottled enamel, a disfiguring brown to black discoloration of the teeth, was being found to be caused by naturally dissolved fluorides in the water supplies of some sections of the country.[6] Margaret and Howard Smith, working at the Arizona Agricultural Experiment Station, even concluded that as little as one part of fluoride per million parts of water was sufficient to produce mottling of children's teeth. Further, aware that orchardists were beginning to use fluorine insecticides, the Smiths analyzed Arizona produce and were alarmed to find that "apples which have been sprayed with barium fluosilicate have shown an average fluorine residue before washing of 5.6 parts per million. . . . A large apple a day alone, there-

[a] Chronic damage was produced by 1.12 mg. arsenic per kg. of rat body weight, as compared, for instance, to a figure of 23.1 mg./kg. for cryolite. These experiments on "chronic" toxicity, however, were carried on for only an eight- to sixteen-week period.

fore, would provide as much fluorine as four glasses of water containing 1 p.p.m. of fluorine."[7]

Various bureaus of the Department of Agriculture had participated actively in the development and testing of fluorine insecticides. The Bureau of Chemistry and Soils, "appreciating the situation which developed in the use of arsenic," even "realized the necessity of having toxicological data available on fluorine compounds before having the Department recommend their use."[8] Floyd DeEds, the Bureau member assigned the fluorine toxicology study, essentially confirmed the work of the Smiths. In an unpublished report submitted early in 1933, DeEds warned: "The substitution of fluorine insecticides for lead and arsenic compounds, the replacement of one inorganic insecticide for another . . . is very apt to be a process of substituting one chronic intoxication problem for another in the field of public health."[9] The DeEds report was immediately submitted to the FDA, and with the findings of the Smiths was to serve as the foundation for the Administration's fluorine spray policy.

The Smith-DeEds work was not as concrete a foundation as the FDA would have liked. It indicated the quantities of dissolved fluoride that could be harmful, and suggested that fluorine spray residues could produce similar injuries, but it was uncertain about the exact amounts of residue that might be dangerous. The question of the solubilities of solid cryolite and barium fluosilicate in gastric juice, of the ease of absorption of fluorine residues, remained to be solved. Thus, while the establishment of a fluorine residue regulation policy was a clear obligation, the absence of any precise toxicological judgment on injurious residue levels imposed a confusing vagueness on that policy. On March 1, 1933 the Department of Agriculture issued a "Notice to Growers and Shippers of Fruits and Vegetables": "The substitution of fluorine compounds for arsenicals in spraying or dusting fruits and vegetables has been urged as a solution of the spray residue problem. There is adequate evidence to establish the deleterious character of certain fluorine compounds and reason to look with suspicion on all such compounds. The presence of fluorine spray residues on fruits and vegetables shipped in interstate or foreign commerce will be regarded as a basis for action under the food and drugs act."[10]

Agriculturalists were left wondering exactly what the FDA meant by this announcement. No definite mention of a tolerance for fluorine residues had been made, but, contrary to established policy with lead arsenate, there had been an implication that no fluorine residue at all would be allowed. Such was the interpretation generally drawn within the agricultural industry; yet when representatives of the industry submitted to FDA officials that the March 1 notice in essence "constituted an absolute embargo . . . against the use of fluorine insecticides," they were told they had misread the announcement.[11] Its intent had been only to discourage farmers from applying the new insecticide with any less care than the old one. The incomplete toxicological record still prevented the setting of an exact tolerance, but it had been thought that since the notice had indicated "that fluorine insecticides had to be used with every bit as much caution as other poisons," farmers would read it to mean "that their application would have to be limited in such a manner that comparable amounts would not be present on fruits and vegetables at the time of shipment." In other words, the FDA had meant to imply that fluorine residues would for the time being be governed the same as arsenic residues. The figure of 0.01 grain per pound had simply been left unspoken, because of the lack of full toxicological support (and also in the hope, no doubt, that the absence of a fixed fluorine tolerance would motivate farmers to hold their residues to the absolute minimum, to be certain of escaping seizures). As much as growers disputed the need for any fluorine residue control, they preferred announced rules to implied ones. They felt more comfortable with something on paper; they demanded a definite fluorine tolerance; and the FDA had to finally consent. Acting on the assumption that fluorine was no more toxic than arsenic, and on the word of Margaret Smith, who in private correspondence had predicted that her experiments "would eventually [show fluorine insecticides] . . . to have about the same chronic toxicity as arsenicals,"[12] the FDA announced a fluorine tolerance of 0.01 grain per pound on June 20, 1933.[13]

It hardly came as a surprise that farmers reacted to the fluorine tolerance much as the old jest would have it that men regard women: unable to live without a definite tolerance before, they now could not live with one. Fluorine residues were even more difficult to remove than lead arsenate, and 0.01 simply seemed an unrealistic tolerance. Several growers' organizations were quick to petition the FDA for a liberalization of the fluorine tolerance,[14] though the clamor for leniency did not build to its peak until 1938. By that time, Washington growers, especially, were beginning to encounter the problem of arsenical poisoning of their orchard soils, and non-arsenical insecticides appeared more desirable than ever. The fluorine compounds were the only effective alternative, but they could never be very convenient so long as the strict 0.01 tolerance remained in force. The files of the Secretary of Agriculture for the year 1938 contain literally hundreds of letters from Washington farmers (and a large number more from orchardists of other states) requesting or, more frequently, insisting that the fluorine tolerance be increased.

These were basically economic appeals, but they were often backed by the weight of what farmers considered scientific argument. In 1937 Simon Marcovitch, the entomologist who had played a leading role in the development of fluorine insecticides, published the results of experiments he had performed at the Tennessee Agricultural Experiment Station. Aided by a chemist and a second entomologist, Marcovitch had fed sodium and calcium fluorides and cryolite to different sets of rats. The rats were weighed weekly and their teeth examined for mottling. It was found that 4 to 7 parts per million of fluorine in the form of cryolite had to be added to their diet to produce "border-line cases" of mottling in fifty percent of the rats. Since the FDA fluorine tolerance was equivalent to 1.4 parts per million, it seemed clear to Marcovitch that "residues on fruits and vegetables sprayed with cryolite have no deleterious or poisonous effect and are not injurious to health," and that the fluorine tolerance could safely be increased tenfold, to 0.1 grain per pound.[15]

Marcovitch's conclusions were as suspect as his motives in carrying out the investigation. He had a clear interest in absolving "his" insecticides, and a critical analysis of his work suggests a certain lack of objectivity. His adoption of fifty percent of the rats injured as the standard for determination of the minimum toxic level, for instance, was contrary to the generally accepted standard of ten percent of the test animals injured. Marcovitch further assumed that the rat is as sensitive to fluorine intoxication as man, and that the toxic fluorine concentration for people would be accepted as identical as that for rats. The Smiths, and other students of fluorine toxicology, had repeatedly stressed, however, that since children's teeth develop much more slowly than those of rats, man should be considered about ten times more susceptible to fluorine poisoning than the rat. Marcovitch and his coauthors maintained that cryolite and barium fluosilicate had lower toxicities than the simpler fluoride salts commonly found in drinking water, yet there was published evidence that differences in toxicity between various fluorine compounds occur only at relatively high levels of fluorine intake, and that the small amounts of fluorine necessary to produce mottling could be easily supplied by any of the fluorine compounds under investigation.[16] After some rather pointless comments on the greater toxicity of dissolved as opposed to solid fluorides, Marcovitch supported his contention that the fluorine tolerance was too strict with the observation that the acute toxicity of arsenic is much higher than for either of the fluorine insecticides, and that the fluorine tolerance should therefore be higher than the arsenic tolerance. The confusion of acute with chronic toxicity was betrayed in other pro-fluorine publications by Marcovitch,[17] but neither this nor the other discrepancies of his work discouraged agriculturalists in pursuit of a tolerance increase. Marcovitch was cited time after time by fruit growers lecturing the FDA on its unenlightened policy.

FDA officials appreciated the objections that could be made to Marcovitch's work, but they were inclined to agree that, as a practical matter, the fluorine tolerance should be reevaluated. Continued shipments of produce in violation of the fluorine tolerance, the mounting complaints by farmers that the tolerance was unfair, and now the publication of the Marcovitch article under the auspices of an institution, the Tennessee Agricultural Experiment Station, with close relations to the Department of Agriculture—all seemed to be drawing the issue of fluorine regulation to a head. When in January of 1938 FDA inspectors uncovered a huge shipment of California grapes contaminated with an average of three times the tolerance for fluorine, Walter Campbell felt that the time for decision had arrived. "The problem of the future policy to be adopted by the Department," he immediately wrote Secretary of Agriculture Henry Wallace, "now confronts us squarely in connection with a consignment of California grapes. . . . This consignment of fruit contains, in our opinion, an amount of fluorine spray residue rendering it adulterated within the meaning of the Food and Drugs Act. . . . However, in the light of those developments indicated in the foregoing statement, it is my judgment that a decision on the question of its seizure or the probable modification of the existing tolerance for fluorine should be made by you."[18]

Wallace accepted Campbell's offer and at once arranged for the appointment of a committee of experts to consider, as the Hunt Committee had done for lead and arsenic, the level of fluorine on produce that would be both safe for consumers and feasible for farmers.[19] In the course of time, on November 14, 1938, the administrative fluorine tolerance was doubled to 0.02 grain per pound.[20]

The new tolerance was a compromise that satisfied no one. Growers considered it still too difficult to meet, and health authorities feared it was too lax to safeguard the public completely. For all their early promise, fluorine insecticides had failed to replace lead arsenate. They had, in fact, not even seriously challenged the older insecticide, and the half decade of argument over fluorine's virtues and vices was really only a side-show to the issue that most fully exercised food officials and farmers alike: the regulation of lead residues.

The threat of a lead tolerance had dangled above the agriculturalists' heads like a sword of Damocles ever since the Salt Lake City Spray Conference. Walter Campbell had warned there that lead was a greater health hazard than arsenic, and implied that its residues would have to be regulated whenever a rapid method of lead analysis could be perfected. The sword fell on February 21, 1933, when the FDA announced its intention to begin seizures of fruit-bearing lead residues in excess of 0.025 grain per pound.[21] This announcement stemmed not only from the recent development of a satisfactory lead analysis, but also from a coincidental sudden increase in lead residues on Northwest fruit. The standard acid-wash method of removing residues had been just as effective for lead as for arsenic and, prior to 1933 lead residues, which were determined periodically by FDA chemists using the older and slower methods of analysis, had not been objectionably high. Late in 1932, however, Yakima and Wenatchee apple growers innocently tested a new washing material, the industrial cleanser tri-sodium phosphate, and marketed apples treated with the new wash. Tri-sodium phosphate, the FDA discovered, removed arsenic as well as the hydrochloric acid wash, but had virtually no effect on lead. Shipments of fruit carrying ten to fifteen times a safe level of lead were found by inspectors armed with the new lead analysis. Thus at the same time that lead residue regulation became possible, it also became urgent.

The first tolerance of 0.025 was, from a toxicologist's viewpoint, rather lenient, being greater than the amount of lead residue theoretically present on an apple carrying the 1933 tolerance of arsenic, and nearly twice the lead tolerance proposed by the Hunt Committee six years earlier. It was intentionally liberal, as had been the original 1927 arsenic tolerance, in order to provide a period of adjustment for farmers before a stricter and safer tolerance was adopted. The FDA's administrative environment had changed somewhat since 1927, however. Its activities were now followed closely by Assistant Secretary of Agriculture Rexford Tugwell, a man with Wileyesque zeal for strict interpretation of the Food and Drugs Act. Tugwell had already taken FDA officials to task for including consideration of the agricultural industry's welfare in their determination of an arsenic tolerance, and when he realized the generosity of the new lead tolerance the Assistant Secretary reacted again.[22] Five weeks after the first lead-tolerance notice on April 1, a second announcement, instigated by Tugwell, lowered the lead tolerance to 0.014 grain per pound, the figure recommended by the Hunt Committee.

Farmers had been distressed by the first announcement, for 0.025 was not easy to achieve. But 0.014 struck them as absurd. There being approximately twice as much lead as arsenic in lead arsenate, the new tolerance would require a cleaning of fruit equivalent to that needed to meet an arsenic tolerance of 0.008, a figure that experience had shown was not consistently attainable. The second lead announcement loosed agricultural furies such as had not been seen before. Speaking in the third person, Tugwell recently recalled the reaction to his tolerance lowering:[23]

"Not many hours after the new tolerance was published, he was visited by two political figures from the State of

Washington: Senator Homer Bone and Lewis B. Schwellenbach, who had unsuccessfully sought the Democratic nomi-
nation for Governor in 1932. . . . Both were Progressives
and he was glad to see them; it would be a conference
among friends. His innocence soon turned to disillusion-
ment. What his friends wanted was to get the new regulations rescinded. . . . Senator Bone and Mr. Schwellenbach
demanded instant reversal of the regulation. Their fellow
Washingtonians were insistent [as witnessed, it might be
added, by the barrage of angry letters that soon hit FDA].
Besides, they wanted to know, how was it that lead poisoning was dangerous now when it had not been in all the
past years. More than that, could he—or could any of the
Food and Drug people—*prove* that anyone had ever died
from the lead in insecticides? Could anyone even prove
that he had been made ill? The whole thing was calculated
to humiliate them in the regard of the folks back home.
It was an acrimonious and inconclusive conversation. And
there were many others."

Among the others were some with Secretary of Agriculture Henry A. Wallace. Tugwell had lowered the lead
tolerance while Wallace was out of town, "only afterward
letting his superior . . . know what he had done." Wallace
no sooner arrived back in Washington than "the Secretary
let his Assistant know that he was hearing from the farm
leaders he was depending on for support on other—and
more important—matters and asked if it was really necessary to allow so minor an issue to interfere with larger
ones."[24] The inescapable nemesis of residue regulation had
reared its head again: the Department of Agriculture could
not permit a peripheral public health concern to interfere
with the legitimate interests of agriculture. On June 20 the
lead tolerance was reset at 0.020 grain per pound.[25]

The whole unhappy episode had perhaps yielded some
fruits of victory for the FDA, since the third tolerance represented a twenty percent reduction of the first, but it had

so angered produce growers and shippers that all future tolerance reductions had to be forced against the most strenuous objections. The FDA set 0.014 as its enforcement goal, and promised that tolerance for the 1934 season, but the goal was never reached. The most that could be managed was a figure of 0.019 for 1934, and of 0.018 for the following three years. In 1938, it will be seen, largely as a consequence of continuing agricultural hostility, the lead tolerance was raised again to its original level.

As with arsenic, the objections to the lead tolerances stemmed not only from the difficulty of meeting them, but just as strongly from the conviction that they were not necessary to protect the public. Farmers were just as skeptical of the dangers of the very small amounts of lead as they were of the hazard of arsenic residues. Since the late 1920s there had been repeated demands from agriculturalists for experimental verification of the claims that spray residues were harmful, demands issued by entomologists and horticulturalists and just plain farmers, all of them confident that the experimental findings would vindicate their spraying practices and free them from regulatory harassment.[26]

FDA personnel shared the desire for more comprehensive experimental data on arsenic and lead toxicity, but because they were certain it would support their policies and make residue enforcement easier. The report of the Hunt Committee had in fact recommended extended investigations of the chronic effects of lead arsenate ingested with fruits and vegetables, but such investigations required considerable sums of money. The FDA was not the most richly endowed federal agency, and not until 1935 was it appropriated the funds and authority sufficient to carry out the recommendation of 1927. On July 1 the Division of Pharmacology was established within the Food and Drug Administration and assigned, as its first task, the determination of the chronic toxicities of arsenic and lead in the form of spray residues. The outline of its research plans drawn up by the Division indicate a determination to provide a thorough basis for understanding the long-term effects of lead and arsenic on the animal organism:

"Groups of animals, chiefly white rats, are to be subjected to varying doses of lead, arsenic, and lead arsenate, respectively, in amounts beginning at the levels of the present tolerances and building up to amounts clearly and relatively promptly toxic. The plan is to examine the animals in this group from every possible angle to get leads as to methods of recognizing toxic effects, and then with these leads to study chronic poisoning in a second group.

"A second group of animals will be given relatively small doses over long periods of time to determine the effects of chronic and repeated injury. . . .

"In these experiments, environmental conditions will be rigidly controlled, especially those of the diet, in order to insure that any physiological or pathological changes which result can fairly be ascribed to the introduction of toxic materials rather than to uncontrolled variations in the experimental conditions. The observations of the groups of experimental animals will consist of the determination of food consumption, growth, activity, appearance, longevity and chemical and pathological studies of organs and tissues. In this manner, intake, storage, and excretion of the metals can be followed in detail. This is of particular importance in view of the apparently authenticated existence of a normal lead stream and of the occurrence of arsenic in various foods. The details of the long-continued chronic experiments will, in part, depend on the findings of the above-described series. It is anticipated that small amounts of lead and arsenic will be given to several generations of animals, observation to be made on such points as rate of growth, age of maturity, longevity, reproductive ability, size, and survival of litters. There are a number of observations which indicate that injury may appear in later generations. From time to time animals will be sacrificed for chemical, pathological, and other studies. . . . After the work is well under way, it is hoped to enlist the cooperation of medical groups and extend the study of both lead and arsenic to the human individual."[27]

To assist the Division in evaluating the data obtained from these experiments, the National Academy of Sciences appointed, at the request of Secretary of Agriculture Wallace,[28] an advisory committee of distinguished university scientists.[b] The collaboration of the two groups uncovered some disquieting information. In a set of articles published in 1938, Division Chief Herbert Calvery and his coworkers reported several experiments in which dosages of arsenic and of lead comparable to those which people might consume as spray residue had produced chronic physiological damage in rats and dogs.[29] These reports were only preliminary, the results of an experimental program intended to run for some time longer, but initial indications were that the present low residue tolerances were justified. Substantiation of these first impressions, however, was not to be obtained, for at the time this group of papers was published, the Division of Pharmacology's spray residue investigation had already been prematurely terminated. It had been exposed to attack by the criticisms of agriculturalists suspicious of animal experimentation, and then brought to an end by the political craftiness of the "Pied Piper of Missouri."

Few farmers had ever been persuaded that poisoning of rats in the laboratory had any relevancy to the question of the harmfulness of sprayed produce for people. Data from experiments on animals was what the fruit and vegetable growers usually referred to as "theoretical knowledge," and their quarrel with the FDA had always been

[b] Chaired by Anton Carlson, the committee included Torald Sollmann, professor of pharmacology, Western Reserve University School of Medicine; Ludvig Hektoen, emeritus professor of pathology, Rush Medical School; Cecil Drinker, professor of physiology, Harvard University School of Public Health; and H. C. Sherman, professor of chemistry, Columbia University.

marked by an insistence on "practical" support for residue regulation. When food officials could point to confirmed evidence of illness among *people* exposed to spray residues, agriculturalists repeatedly implied, then the FDA might expect full cooperation with its residue control program. In the meantime, growers would not allow their personal experiences—their failure to be killed or sickened by their produce—to be contradicted by rats.

The calls for residue toxicity studies on human populations became conspicuously numerous during the early 1930s, prompted by continuing embarrassments in the export apple market. England had objected to American apples on several occasions since 1926, and by 1930 other European countries were beginning to take similar exception. The situation finally deteriorated to the point when, in the summer of 1931, the secretary of the International Apple Association frantically informed the FDA's Campbell:

"Here is a nice mess. Today I received a letter from one of our members who is now abroad reading as follows: 'ARSENICAL SITUATION: I shall cable you in a day or two of this which is being reported by Department of Commerce men to Washington. In Czecho-Slovakia and Austria, newspaper articles and radio talks are telling the public that American apples are poisonous, and Poland has actually prohibited the entry of American apples. The Department of Agriculture should, thru [sic] the Associated Press, issue strong denials of any amount of spray residue harmful to health—should explain about analysis . . . wiping and washing, and should immediately acquaint all Department of Commerce men in Europe of our caution and protective measures. We shall feel the loss of Poland if they don't reverse their stand and the International Apple Association and the Senators and Representatives from apple states should endeavor to get our Government to impress upon foreign governments that no danger exists. This exists primarily in Czecho-Slovakia and Austria, is not appar-

226

ent yet in Hungary, but has been the subject of radio talks in France. Poland, of course, is the worst so far.' "[30]

An interesting handwritten postscript was appended to the letter, an observation that "if this thing keeps on, you will have to go to 0.01 before next year in order to protect our foreign commerce."

The inference that apple growers could somehow adjust their practices to meet the world tolerance when their financial interests were threatened, but not if the health of the public were the only risk, is perhaps not an entirely fair one, for orchardists did not accept that the public was endangered by their product. They were certain that if competent scientists would investigate the effects of residues on people, residue regulations would be exposed as unnecessary, and costly, foolishness. A conference of Washington growers that met in Spokane in the spring of 1933 exemplified this belief of the agricultural industry, the faith that "an examination of a few thousand persons who are exposed, as they are in the Wenatchee and Yakima regions, could be completed in a few months. . . . If that research were made and the results of such investigation published in some journal of international circulation . . . it might do much to eliminate the uneasiness in some foreign countries as well as [in] our own country."[31]

The Spokane plan was a typical proposal of farmers threatened with the loss of buyers at home and abroad, and a plan that assumed increased attractiveness after the Division of Pharmacology was established. No sooner, in fact, had the Division announced its plans for toxicological studies on laboratory animals than the American Pomological Society initiated a drive to replace what its members considered inappropriate experimentation with an investigation of human susceptibility to residues.[32] Its Committee on Research on Spray Residues presented the Society's plan to the Food and Drug Administration and to the U.S. Public Health Service (at that time part of the Treasury Department), but it was ignored by both. The plan seemed superfluous while the Division of Pharmacology was already pursuing a project that intended to examine human response to spray residues once sufficient animal data was obtained. The impatient Pomological Society discovered that it lacked the political clout required to reorder governmental research programs, but the needed ally was soon found in the Honorable Clarence Cannon.

Clarence Cannon represented the rural Ninth District of Missouri in the House of Representatives for more than forty years, from 1923 until his death in 1964. For nearly all of the second half of that period, he ruled as the chairman of the House Committee on Appropriations, and earned a reputation as the no-nonsense, budget-slashing "watchdog of the national purse strings."[33] The one legislative mission possibly dearer to his heart than the saving of taxpayers' dollars was the improvement of the quality of life in the countryside, and Cannon was throughout his career a tireless advocate of soil conservation, reclamation and flood control projects, rural electrification, and parity payments. During congressional recesses, when back home in Missouri, Cannon was also an apple grower and, like most apple growers, he protected his fruit with lead arsenate and ate the fruit with no fear of being poisoned. It was his opinion, and one that he frequently shared with the FDA, that the "whole spray residue campaign has reached the point of absurdity and is without foundation of any fact or scientific reason whatever."[34]

In 1935 the appropriation bill for the Department of Agriculture, including the FDA, had to be examined by a House appropriations subcommittee of which Clarence Cannon was a member.[35] Since one-third of the FDA's expenditures for enforcement of the Food and Drugs Act was at this time being applied to residue control, Chief Campbell might have expected to face a trying interview when called before the subcommittee to defend his Administration's funding requests. Mr. Cannon did not disappoint him. At earlier sessions, Cannon had questioned the heads of other Department of Agriculture bureaus about the dangers of spray residues, and these had all indicated that Campbell was the proper man to provide such information. When the FDA's turn before the subcommittee arrived, Campbell was first quizzed about the general operations of the Administration, and then directly challenged by Cannon to cite specific instances of injury done to people by sprayed produce. Campbell had come prepared with a catalogue of cases of acute poisonings, prepared from FDA files, and he presented these with a warning that the chronic danger of residues was even greater than the acute one.

To Campbell's FDA colleagues witnessing the proceedings, Cannon seemed thoroughly refuted. As a Campbell assistant was later to recall, however, the Representative from Missouri had only begun to fight:

"When the official printed record of the hearings consisting of 1,693 pages for the Department of Agriculture as a whole reached my desk a few days later, I turned at once to that part relating to the Food and Drug Administration. I was eager to see if Campbell's words were as convincing in print as when spoken to the committee. To my amazement the printed record contained nothing of his statement about spray residue! Not a single word of his evidence on the danger of lead arsenate on apples had been printed. Someone had deleted every word he said on the subject. Even more amazing was the fact that Cannon's challenges to other bureau chiefs to show a single instance of the harmful effects of spray residue were printed. Also printed were the replies of the other bureau chiefs who stated that Campbell was the man to whom Cannon's questions should be directed. All fingers pointed to Campbell, but when one turned to this testimony as printed, it appeared he had failed to accept Cannon's challenge, the implication being that he was unable to furnish any evidence in support of the spray residue program."[36]

Cannon's suspected role in the deletion of Campbell's

testimony could not be proved, but an appeal was made to the chairman of the full Committee on Appropriations, and the FDA was assured that any and all comments made at future hearings would be included in the record.

A man of Cannon's political experience was versed in an assortment of tactics for achieving legislative ends. When FDA personnel read through the Department of Agriculture appropriation bill for the fiscal year beginning July 1, 1937, they were stunned again, this time by a clause requiring that "no part of the funds appropriated by this Act shall be used for laboratory investigations to determine the possibly harmful effects on human beings of spray insecticides on fruits and vegetables."[37] This effective liquidation of the Division of Pharmacology's spray residue study was the work of Cannon, now chairman of the subcommittee in charge of agricultural appropriations, and indicated he shared the belief that toxicity experiments took too much time and gave misleading results. People had to be studied, and soon, if agriculture were to be saved, so Cannon balanced his deduction from the FDA budget with an addition to that of the Public Health Service. After discussing with the Health Service the feasibility of a field study of the effects of sprays and residues on human beings, Cannon testified in support of funding such a project at the hearings on the Treasury Department appropriation bill. "There is probably no other investigation," he told his fellow Representatives, "which could be of such immediate, direct and comprehensive interest to the people of the United States."[38] On July 1, 1937, the U.S. Public Health Service was directed to carry out a study "to determine the possibly harmful effects on human beings of spray insecticides on fruits and vegetables."[39]

The Waterman specter of political interference with residue regulation had materialized. The Division of Pharmacology had felt that it was only a year away from conclusive demonstration of the injuriousness of ordinary spray residues, but by governmental order it had been forced to destroy all its experimental animals (some 5,000 rats) and halt its investigation. The preliminary findings were nevertheless published, prefaced by a history of the study and its sudden demise that only partly disguised the authors' rancor. Supporters of strict residue regulation from outside the civil service were more openly angry with Cannon. A cartoonist pictured him as a "Pied Piper of Missouri" leading the Division of Pharmacology's experimental rats to their destruction,[40] while the ever pugnacious Schlink even compared the killing of the test animals to Hitler's burning of liberal books.[41]

Greater dismay was to follow the announcements of the Public Health Service's findings. The transfer of spray-residue investigations from the FDA to the Health Service did not at first seem entirely ill-advised, since the latter agency had considerable experience with epidemiological studies of metal intoxication (particularly of lead poisoning), and, being outside the Department of Agriculture, might even be expected to make a more impartial investigation. The program designed by the Health Service and begun early in 1938 was a field study of the health of more than 1,200 people (apple growers and packers and their families) in the area around Wenatchee, Washington. It was felt that if any population group in the country had been adversely affected by spray residues, it should be those who had worked with sprays and consumed sprayed produce in large quantities for as long as forty years. Grouped by sex, age, and extent of exposure to lead arsenate, the Wenatchee residents were to be given periodic physical examinations as checks for injury due to lead or arsenic.

Concurrently with the field study, laboratory investigations were being conducted on men and animals. The Public Health Service's first report was released in mid-1938, and pertained to these laboratory experiments: "One hundred milligrams of lead arsenate were ingested by two individuals over a period of ten days while on a controlled diet. The degree of absorption, path of excretion, and toxicity of this dosage were evaluated. . . . While the lead arsenate was completely broken down in the body, no untoward effects on the well-being of these two individuals attributable to this quantity were noted. The greater part of the lead and arsenic derived from the ingested lead arsenate was directly recovered, and it was found that the lead was excreted with the feces and that the arsenic was excreted in the urine."[42]

Anton Carlson, a long-time opponent of lead arsenate and the chairman of the advisory committee to the Division of Pharmacology residue study, characterized one set of reactions to the Public Health Service report by ridiculing the suggestion that ten days of experimentation on two subjects was adequate to determine anything about the chronic effects of lead arsenate.[43] Clarence Cannon represented the other response to the Public Health Service announcement. Shortly after its publication, the congressman wrote the Department of Agriculture from his home in Missouri suggesting that, in the light of this new evidence, residue tolerances might be raised.[44] The Cannon letter was referred to the FDA, which replied that the Public Health Service had not specifically recommended any tolerance increase. Cannon's supply of patience was exhausted:

"Have your letter," he fired back, "and am surprised to note that while the report from the Public Health Service shows no need for a low tolerance, the Department takes refuge in the statement that the report includes no conclusion and no recommendation for any change. The report positively does definitely include a conclusion of no ill effect of the injection of spray residue. It states its conclusion unequivocally and in so many words. Of course, it is not the function of the Public Health Service to recommend a 'change.' That is the function of the Secretary of Agriculture . . . the Department [should] make amends for the campaign of terrorism they have been carrying on and the inestimable damage to one of the great agricultural industries and to the country as a whole through the alarm resulting from these newspaper and radio and bulletin releases by the Department of Agriculture which have so frightened the mothers of the nation that they have deprived their children of one of their most valuable articles of diet. It is high time the Department of Agriculture desisted from this witch-burning crusade and got down to facts and to the discharge of obligations which it owes to the government and to the consuming public."[45]

Though the FDA was not bound to conform to Public Health Service advice, with the political handwriting so clearly on the wall it seemed advisable at least to seek the Service's opinion. It was in fact immediately sought, and followed, and on September 19, exactly one week after the writing of the Cannon letter, a new lead tolerance of 0.025 was announced.[46]

Liberalization of the lead tolerance was by no means entirely a Cannon coup, for the Department of Agriculture and the FDA were swayed as well by the hundreds of letters from orchardists informed of the Public Health Service report through their growers' organizations. It had been demonstrated many times before that the FDA could not stand firm when agricultural opinion mobilized against it, and, no matter how dubious and preliminary, the Health Service findings offered a distinct rallying point for mobilization. But even their latest triumph did not satisfy farmers for long. As the leader of the Yakima area growers pointed out, the tolerance increase would not be as helpful as had been hoped, since the arsenic tolerance was still at 0.01 and careful residue removal would have to be continued.[47]

The hollowness of their victory quickly became apparent, and farmers began to bombard the FDA anew. A popular theme of the latest barrage of complaints was the demoralization of America's rural citizens by the Food and Drug Administration's "discriminatory rulings." A Wenatchee grower, anticipating his country's entrance into the war against Hitler, asked, "What may we expect to happen to the sons of farmers when it comes time to fight? Will the sons of tobacco magnates, and brewery and distillery tycoons, have the same advantage over the sons of fruit growers in the draft that their fathers have over fruit growers in the sale and distribution of their products? Will our boys fight under different flags? Or will boys from farms grow so discouraged by discrimination that they will say, 'Oh, hell, what's the use!' "[48]

Another letter from Washington warned quite bluntly that FDA tyranny "is making communists of farmers who constitute the last bulwark of that kind of conservatism which once constituted real American democracy."[49]

These were hardly the only agriculturalists to detect "subversive influences . . . inside our own government," and this situation was made all the more intolerable by the fact that these influences were draining the farmer's purse as well as his democratic spirit. For more than a decade, growers argued, they had borne the cost of unnecessary residue removal, and they could not continue under the burden much longer. They demanded relief, and some even insisted on restitution. Only eight months after liberalization of the lead tolerance, the Master of the Washington State Grange, an organization that had lobbied vigorously for that liberalization in the belief that it would salvage the farmer's declining fortunes, presented Franklin Roosevelt with a petition bearing the signatures of over 5,000 Washington orchardists. The petition called the President's attention to thirty-seven facts illustrative of the fruit grower's plight, and concluded with the observation "that the United States Department of Agriculture has admitted its error with reference to the matter by recently relaxing its lead tolerance 39 percent and its fluorine tolerance 100 percent. In view of these facts and the tragic losses to the apple growers caused by mistakes of the United States Department of Agriculture, an agency created for the benefit of farmers, we the undersigned, do hereby petition the government for compensation for our losses and very respect-

234

fully request that an amount of not less than $100,000,000 be made available for this purpose at an early date; and further, that the tolerance for spray residues be set at such a figure as will conserve the health of the consumer and at the same time enable the growers to wash their apples on their own farms without damage to the fruit."[50]

The petition's request was absurd, especially coming during a period of still painful national economic recovery, yet its timing makes it all the more significant as an indicator of agricultural recalcitrance. The traditional policy of educational rehabilitation instead of immediate punishment of food adulterators had been applied to residue regulation for twenty years, and many farmers still considered the control program objectionable and cooperated with it only under protest. The ghost of Harvey Wiley was no doubt laughing and reminding the FDA, "I told you so."

Through most of the 1930s, Food and Drug Administration personnel were too busy laying the groundwork for a new, broader, and stronger enforcement program to be distracted by any spectral derision of the old program. The 1933 decision to replace the obsolete Food and Drugs Act with a sounder model was immediately followed by a gust of brain-storming and legislative drafting, by FDA officials and legal advisers, that shortly produced a bill containing "everything we thought the interest of the consumer required."[51] It was indeed a bill of broad scope, one that sought to extend FDA protective rule into much hitherto ungoverned area. The original draft provided for the regulation of cosmetic products and of mechanical devices sold as therapeutic agents; it intended to prohibit false and misleading advertising of foods, drugs, and cosmetics and the sale of remedies that might be injurious even though sold without false or fraudulent curative claims; it allowed for the establishment of legal definitions and standards for foods, and for the setting of official tolerance levels for poisons unavoidably included in food products.

These provisions, supplemented by the bill's lesser proposals, represented such a radical investment of power that legislators willing to support the FDA's claim on Capitol Hill were difficult to find. Only after "Dr. Tugwell and Mr. Campbell literally peddled the thing up and down the halls of Congress"[52] did a Senator from New York, Royal Copeland, step forth and offer to assume the burden. Copeland was a homeopathic physician, erstwhile health commissioner of New York City, and a lawmaker who had become appreciative of the FDA's responsibilities and legal limitations during congressional hearings on adulterated drugs several years earlier. He was not a Senator with a reputation as a fighter, but Copeland at once found himself, after his introduction of the FDA bill on June 12, involved in the fight of his life.

Drug and cosmetic manufacturers, and to a lesser extent food processors, were horrified by the bill's provisions, and damned it as potentially "the greatest legislative crime in history."[53] Manufacturers were equally repelled by the bill's sponsorship. It was precisely, they pointed out, what might have been expected from that "cocksure brain-truster Prof. Rexford G. Tugwell."[54] The Assistant Secretary of Agriculture, so influential in initiating the campaign for new food and drug legislation, was not one of the best-loved New Dealers.[c] He had a personality that, as his superior in the Department of Agriculture described it, "exhibits disdain,"[55] and was an advocate of decidedly liberal economic views at a time when Bolshevism seemed a palpable menace. Any object branded with the name of "Rex

[c] But he was, judging from the results of a beauty contest held by a Washington newspaper using its female subscribers as judges, the best-looking New Dealer. Cited by Russell Lord, *The Wallaces of Iowa*, p. 348.

the Red" invited attack, and the "Tugwell Bill" introduced to Congress by Senator Copeland was hardly overlooked. "When they say anything bad about the bill," Copeland observed soon after hearings on it began, "they call it the Tugwell Bill; when they say something nice, they name it the Copeland Bill."[56]

This was too hopeful a generalization, for Copeland himself was being assaulted as fiercely from the left as was Tugwell from the right. America's top muckrakers and consumer champions, Arthur Kallett and F. J. Schlink, denounced the bill as weak and inadequate, and lambasted the selection of Copeland as sponsor because of his appearances on radio shows associated with offending drug products.[57] Copeland regularly delivered popular health talks on broadcasts sponsored by such products as Fleishmann's Yeast, Phillips Milk of Magnesia, and Pluto Water, preparations whose manufacturers might expect to suffer if the Copeland bill were passed. The Senator's acceptance of pay from shows sponsored by proprietary remedies, Kallett and Schlink charged, was a conflict of interest certain to hinder the enactment of effective drug legislation.

Copeland was hardly a legislative saboteur, but he was political realist enough to know that industrial opposition to the original draft of the bill was much too strong to be overcome. A revised and less ambitious version would have to be prepared and submitted to Congress if any new law at all were to be obtained. Walter Campbell reluctantly agreed, and although Tugwell declined to identify himself with any compromise, Senator Copeland spent the closing weeks of 1933 modifying the "Tugwell-Copeland" bill into a purely Copeland proposal. Dropped into the hopper on January 4, this toned-down version, and the several Copeland revisions that followed, became the object of a protracted political struggle reminiscent of the one that preceded the 1906 Act, and that similarly might have continued some time longer had not scandal intervened.

Elixir Sulfanilamide was *The Jungle* of 1937. Sulfanilamide, first of the group of "wonder drugs" that excited such medical and popular enthusiasm during the 1930s, was originally administered as tablets or capsules, but in the late summer of 1937 the Massengill Company of Bristol, Tennessee, began marketing a liquid form of the drug. Their elixir employed diethylene glyco as solvent, tests having shown it to have the most agreeable smell and flavor of those liquids able to dissolve sulfanilamide. The Massengill chemist, however, had neglected to test the solvent's toxicity, and by the time the public had finished testing it for him, more than one hundred people had been killed by Elixir Sulfanilamide, their kidneys and livers irreparably damaged by diethylene glycol.

To make these inexcusable tragedies even more loathsome was the fact that legally there was little that could be done to punish the elixir's manufacturer. As the FDA and other supporters of the Copeland bill had been pointing out for years, the Food and Drugs Act did not prohibit inclusion of toxic ingredients in medicinal preparations. It merely required that the presence of any of a small group of recognized poisons be acknowledged on the label. Diethylene glycol was not one of the poisons required for labelling, and the only law the Massengill Company had broken, by as careless a chemical oversight as the use of toxic solvent, was the provision against misbranding. Legally, any drug product carrying the name "elixir" had to contain alcohol, and Massengill's sulfanilamide did not. Samuel Massengill was tried and convicted on 174 counts of misbranding, and fined $150 for each, but his punishment of more than $26,000 clearly fell short of fitting the crime.

The Elixir Sulfanilamide disaster impressed upon the public, as no amount of muckraking propaganda could have done, that they were in fact helpless guinea pigs and would remain so until they provided themselves with adequate legislative protection. Largely due to the national revulsion created by the sulfanilamide deaths, the Copeland bill finally passed into law in June 1938. Included in this Federal Food, Drug, and Cosmetic Act was a provision requiring that manufacturers of new drugs file with the Secretary of Agriculture a statement of experimental tests demonstrating the remedy to be safe before introducing it into interstate commerce. Four days after its passage by Congress, and a week before the Act was signed by President Roosevelt, Royal Copeland collapsed and died "of a condition rendered acute by overwork."[58]

Yet even after the sulfanilamide tragedy, there had remained opposition to certain provisions of the Copeland bill, compromising revisions had to be made, and there was one major issue that had to be resolved before the bill could become an act. Fruit growers had followed the fortunes of the Copeland bill with a wary eye, for it carried a provision that threatened to make the spray residue situation worse than ever. This was the proposal to grant to the Secretary of Agriculture the authority to establish legal tolerances for poisons added to food products. The drafters of the provisions had had sprayed produce in mind when drawing it up, since residue regulation was the chief consumer of FDA time and money during the mid-1930s, and one of the greatest nuisances to this regulation was the necessity of defending seizures of sprayed produce in court. Armed with official tolerances, the FDA would not have to waste its energies trying to persuade a jury that one-hundredth of a grain of arsenic might be harmful, but could concentrate its courtroom activity on the simple demonstration that the seized product carried residues in excess of the established tolerance.

The FDA's gain would obviously be the agriculturalists' loss, but it was perhaps not anticipated that growers would regard the loss so seriously. Coming on the heels of the announcement of the hated administrative tolerance for lead, the official tolerance proposal struck farmers as another step, and a large one, toward confiscation of their freedom. The chairman of the Northwest Spray Residue Committee, a Washington apple grower, typified the fruit industry's attitude toward the Tugwell-Copeland bill when he analyzed the tolerance provision for his fellow orchardists as a form of martial law investing the Secretary of Agriculture with dictatorial powers. It would serve, he promised, "to destroy the rights in criminal prosecutions which have been written into the Anglo-Saxon system of jurisprudence since the time of the Magna Charta."[59]

The new tolerance proposal had to be defeated, and to take up the sword in the farmers' behalf came the International Apple Association. Formerly the International Apple Shippers' Association, the organization had in the past performed some liaison duty for the Bureau of Chemistry and the FDA, serving as a friendly and trusted source of information for growers on the need for residue controls. It could not, however, cooperate with a Food and Drug Administration that proposed to set its controls beyond the reach of judicial process, and soon the figure of the Apple Association's Samuel Fraser, "with his trailing walrus mustache and heavily laden brief case," became "a familiar sight on The Hill in Washington . . . [trudging] up and down the corridors of the Capitol or in and out of the Office Buildings. . . . What his mysterious errand may be he never divulges to outsiders."[60] His "mysterious errand" was to convince legislators of the injustice of the official tolerance provision of the Copeland bill.

Fraser began to offer serious public criticism of the tolerance proposal at the 1935 Senate hearings on the Copeland bill.[61] He hinted then that he had in mind some amendments to rectify the provision, but among so many other issues competing for attention at these early hearings, Fraser's comments passed unnoticed. He was relegated to wandering about the Capitol and congressional office buildings again until 1938, when a fertile opportunity for amendment of the tolerance provision at last presented itself. In March of 1937 the latest version of S.5, the Copeland bill, had been passed by the Senate and sent to the House of Representatives. There it became locked in committee for many months, in a committee chaired by Representative Lea of California, a leading fruit-producing state. As one controversial aspect of the bill after another was threshed by the committee, and the decisions of what to keep and what to discard were made, the number of issues for consideration dwindled until only the official tolerance provision remained. Samuel Fraser now received his day in court, and Representative Lea lent a sympathetic ear.

One of the long-standing provisions of the Copeland bill was that any regulation (including poison tolerances) issued by the Secretary of Agriculture would be established only after affected parties had been given opportunity to present their views on the subject at a public hearing. When S.5 finally emerged from the House committee in April 1938, it contained a new section providing that within ninety days after issuance of a regulation by the Secretary, suits could be filed in any federal district court enjoining the Secretary from enforcing the regulation. The court would have authority to collect new evidence relating to the regulation and to pass on its validity. In principle, this judicial review clause would satisfy the fruit growers' demands that the Secretary not be given dictatorial power, that he not be allowed uncontested freedom to set spray-residue tolerances where he wished. More importantly, in practice it meant that the enforcement of new residue tolerances could be at least temporarily frustrated. There were more than eighty district courts in the country, and the filing of anti-tolerance suits in one court after the other could postpone effective institution of a tolerance for years.[62]

The FDA at once foresaw the impossibility of residue control should the judicial review clause become law. Speaking for the Administration, Secretary of Agriculture Henry Wallace publicly asserted that legalization of the clause would "amount to a practical nullification of the substantial provisions of the bill," and proclaimed that, "It is the department's considered judgment that it would be better to continue the old law in effect than to enact S.5 with this provision."[63] There was a great deal more criticism of the clause from consumer organizations, but Lea remained adamant, probably, it was suspected by FDA personnel, because of the "forthcoming campaign in which the support of fruit growers will be a substantial element in his success at the polls."[64] He did, however, offer persuasive arguments for judicial review, presenting it as a choice between "government by edict" or "government by orderly procedure, a government under which the citizens shall have a right to be heard."[65] The Lea version of the Copeland bill was passed by the House.

In the ensuing conference between Senators and Representatives called to reconcile the differences between the two food and drug bills they had passed, judicial review again became the chief point of disagreement. The struggle over the provision was so spirited, in fact, that it threatened to end in stalemate until the President, at the continued prodding of Secretary Wallace, let drop the implication that he would veto any bill containing the House judicial review amendment. In these circumstances, a compromise court review section was quickly agreed upon. The bill that issued from the conference to be signed into law ten days later, contained a Section 701 (f) which specified that "In a case of actual controversy as to the validity of any order under subsection (3) [any regulation established by the Secretary of Agriculture], any person who will be adversely affected by such order if placed in effect may at any time prior to the ninetieth day after such order is issued file a petition with the Circuit Court of Appeals of the United States for the circuit wherein such person resides or has his principal place of business, for a judicial review of such order."

Fruit growers interpreted Sec. 701 (f) as a victory,[66] and so did the FDA. The Administration's tolerances could still be challenged, but only in ten circuit courts instead of the originally proposed eighty-five district courts, and only by persons living within the circuit in which the petition was filed. The institution of tolerances would be much less cumbersome than if the International Apple Association had had its way and, once instituted, tolerances would no longer be subject to jury review after every contested seizure. On balance, the Food, Drug, and Cosmetic Act granted the FDA freer control over spray residues than it had had in the past.

What added power it received was in a sense unneeded, for by 1938 the FDA considered the problem of arsenic, lead, and fluorine residues to be declining. During the fiscal years from 1933 through 1937 the Administration had allotted approximately thirty percent of its working time to residue regulation.[67] In 1938 that percentage fell abruptly to twenty-three, due to the diversion of activity toward Elixir Sulfanilamide seizures, but in 1939 the time allotment figure dropped still further, to about twenty percent. Campbell announced that the decrease in residue regulation time by fully one-third in two years represented "the continued improvement in the spray-residue situation."[68] He had already declared in 1936 that, "It may be properly reiterated that interstate traffic in sprayed fruits and vegetables no longer presents a public-health hazard."[69]

A tabulation of the annual number of seizures of sprayed produce from 1933 through the 1940s indicates that Campbell was a little premature in his proclamation.

Year	Number of Seizures	Year	Number of Seizures
1933	241	1942	16
1934	58	1943	4
1935	338	1944	20
1936	146	1945	32
1937	125	1946	20
1938	297	1947	11
1939	91	1948	1
1940	98	1949	2
1941	4		

A major permanent decrease in the number of seizures did not occur until the 1939 season, and the most significant reduction not until 1941. The reason given for the decline in residue violations during the late 1930s was "the increasing concern with which the industry and State authorities are viewing the spray-residue problem and are adopting corrective measures at the source."[70] No doubt more conscientious efforts to secure unobjectionable fruit were being made at the local and state levels, but the sudden near-disappearance of residue infractions in 1941, and the continued low number of seizures throughout the decade, were the result primarily of two other developments.

The responsibility for determining the toxic effects of lead arsenate, it will be remembered, had been transferred by congressional fiat from the Food and Drug Administration to the Public Health Service. A preliminary report had been published by the Health Service in 1938 and been used as the basis of the lead tolerance increase. The full project, a study of the residents of Wenatchee, Washington to determine what effects, if any, they had suffered from their long exposure to lead arsenate, was not completed until 1940. Its findings, reported the following year, were that of 1,231 people studied closely for the past three years, only seven, all orchardists who regularly engaged in spraying, had shown "a *combination* of clinical and laboratory findings directly referable to the absorption of lead arsenate. Some physicians may interpret these cases as minimal lead arsenate intoxication."[71]

The toxicological impasse that had prevented agreement on the danger of residues before was encountered again. The subjects of the Wenatchee study had been given exceptionally thorough physical examinations, including histories of their exposure to lead arsenate, chemical analyses of blood, urine, and feces, and even fluoroscopic examinations of the skeleton in some cases. Yet the study could not hope to escape criticism, because by its nature it could detect only acute, sub-acute, and advanced chronic intoxication. As A. J. Carlson had warned at the outset of the Public Health Service investigation, "We must be sure that lead and arsenic produce no injury antecedent to frank and easily recognizable poisoning. The usual medical examination for health cannot tell us that."[72]

The only thing that could indicate sub-clinical damage was extended experimentation with animals whose internal organs could be examined, the kind of experiments the FDA's Division of Pharmacology had begun in the mid-1930s. But those experiments had been ended, and the Public Health Service study authorized and funded, precisely because the agricultural industry refused to accept data obtained from rats and guinea pigs as applicable to people. A risk had to be taken one way or another: either the assumption that animal data was valid, and that the original low tolerances were therefore justified, had to be accepted and farmers forced to take the risk of spending more money to clean their fruit than was actually necessary for the public's protection; or, the assumption that clinical examinations of people gave sufficient toxicological information could be accepted, in which case the public would incur the risk of being slowly poisoned. Had public health considerations been the only factors involved in the decision, the first alternative would have been quickly selected, but the contribution of economic pressures from the agricultural industry necessarily weighted consideration of the problem toward the second solution. In 1940 the FDA was removed from the Department of Agriculture to the newly created Federal Security Agency, but even this long-desired administrative transfer did not bring complete relief from agricultural influence. The financial abilities of farmers still had to be considered alongside the toxicological record, and any objections that might be made to the Public Health Service investigation were less moving than the continued complaints of fruit growers. On August 10, 1940, soon after completion of the Wenatchee study, the Administrator of the Federal Security Agency announced an in-

crease of the arsenic tolerance to 0.025 grain per pound and
of the lead tolerance to 0.05 grain per pound.[73]

Reaction to this turn of events was, of course, mixed.
Carlson wired the editors of Consumers' Union Reports
that he had "reviewed all work on lead and arsenic poisoning by the Public Health Service and in my judgment that
work furnishes no scientific basis for permitting more lead
and arsenic on apples and pears."[74] Another scientist, the
horticulturalist T. J. Talbert, disagreed, rhapsodizing that
"truth has finally been triumphant over superstition, con-
jecture, and hearsay evidence."[75] But even as the two sides
sniped at each other, each was girding for an anticipated
full-scale battle. The new tolerances were still only administrative tolerances, figures that would have to be defended
in court if a seizure were contested. Under the provisions
of the Food, Drug, and Cosmetic Act, a tolerance could not
be made official until a public hearing had been held, and
it was at this public hearing that consumer advocates and
agriculturalists expected to clash and settle the tolerance
question for good. The conflict, however, kept being postponed. The FDA's original plan had been to continue
under the old system of administrative tolerances until the
final report of the Public Health Service could be published, after which a hearing would be announced.[76] But,
as Paul Dunbar, Campbell's successor as Commissioner of
Food and Drugs, later related: "the war intervened and
made it impossible for us to hold the necessary hearings to
establish tolerances. . . . Following the termination of the
war, we began preparations to reopen the hearings and set
up tolerances under the law for a variety of insecticides
which were coming into extensive use. Then we were confronted with a difficult situation. During the war a large
number of new and very potent insecticides had been developed. Little was known about their toxicity either to
the person who applied the sprays or to the consumer who
ate the finished food product. In several cases we didn't
even have methods for accurate estimation of the residue

spray left on or absorbed by the food product. . . . We knew
too little about many of these insecticides to hold hearings
and establish safe tolerances."[77]

The hearings were finally held in 1950, but offered, so
far as lead arsenate was concerned, only a rehash of evidence that had been familiar to the FDA for years. The
official tolerances for lead and arsenic announced after the
hearing, though expressed in new units, were essentially
the same as the administrative tolerances of 1940. Arsenic
was set at 3.5 parts per million, lead at 7 parts per million.[78]
In the meantime, the liberalized administrative tolerances
had been so much easier to meet that the number of seizures of sprayed produce stayed unusually low throughout
the 1940s. The decline did not, however, mean, as Campbell had wishfully thought, that spray residues would no
longer be a hazard; rather, arsenic and lead were simply
being replaced by residues from insecticides whose dangers
were even less clear.

*The possibilities of DDT are sufficient to
stir the most sluggish imagination. . . . In my
opinion it is the War's greatest contribution
to the future health of the world.*

—Brigadier General James Simmons,
Saturday Evening Post, Jan. 6, 1945

Epilogue

ROSEATE EXPECTATION flourished among the early commentators on this "wonder insecticide of World War II."[1] The insecticide had been developed during the late 1930s by Paul Muller, of the Swiss firm of J. R. Geigy,[2] and had first demonstrated its potency in 1939 against, ironically, the Colorado potato beetle (long a European pest by that time). The first American tests on DDT were not possible until 1942, but the product scored so impressively in these that virtually all available supplies were immediately appropriated by the U.S. Army. Pyrethrum and rotenone, the insecticides normally used against disease-carrying insects, were in short supply during the war, and military success depended in no small measure on the army's ability to subdue louse-borne typhus in Europe and mosquito-borne malaria in the Pacific. DDT served in both theatres, though most spectacularly in the Italian campaign. When typhus was introduced into Naples in December 1943 dusting stations were established throughout the city, and over the next three months more than two million civilians reported to the stations to have DDT powder sprayed under all their clothing. With the insecticide being used so freely as even to replace rice at wedding ceremonies, the incipient epi-

248

demic was quelled with no loss of American lives. Soldiers in the Pacific were at the same time becoming acquainted with DDT in the form of the "Flying Flit Gun":[3]

"Avenger torpedo bombers, equipped with nozzles for spreading a spray of DDT and Diesel oil, have winged low over Pacific islands, blotting out almost entire insect populations. As a result, in one island recently wrested by the marines from the Japs, not a single case of insect-borne disease has been reported. . . . Beginning less than a day after the captured airstrip was put into operation, Marine pilots began systematically spraying every square yard of the island, thus giving the Nips a few insect-free days before they were mopped up."

When the war ended, DDT was given a hero's welcome, hailed throughout the land as "Killer of Killers" and "the atomic bomb of the insect world."[4] More importantly, it was bought. "The army's new insect-killer has nearly every householder pawing the ground in eagerness," popular magazines reported, and "people . . . are raiding the stores for every can that shows its top above the counter."[5] Farmers most of all were rushing after the new insecticide. Food and Drug Commissioner Paul Dunbar explained the record low number of residue seizures for 1948, a total of one, with the observation that, "In general, arsenic and fluorine sprays have been replaced by DDT and other newer chemicals."[a]

[a] *Federal Food, Drug, and Cosmetic Law, Administrative Reports, 1907-1949*, Washington, 1951, p. 1,346. Arsenicals nevertheless continued in use for certain pesticidal purposes, but on an embarrassingly small scale. In 1945, on the eve of the DDT takeover, more than thirty-five tons of lead arsenate were manufactured for agricultural use. By 1960, barely five tons of the product were being produced (J. C. Headly and J. N. Lewis, *The Pesticide Problem: An Economic Approach to Public Policy*, Baltimore, 1967, p. 7). The "other newer chemicals" replacing arsenic and fluorine during the post-war years are too numerous to list. Among the more popular were aldrin, BHC, chlordane, dieldrin, lindane, malathion, and parathion. Descriptions of these and many more may be found in *Insects, the United States Department of Agriculture Yearbook of Agriculture*, 1952, pp. 748-750.

What few warnings there were that the new insecticide might prove as dangerous as the old were generally dismissed as hysteria. *Reader's Digest* reflected the attitude of most of the popular literature when it denounced the "fantastic myths [which] have been built up concerning DDT's . . . deadliness to men and women, to children, to pets."[6] More authoritative scientific opinion concurred. The author of a 1945 article discussing the first known case of DDT poisoning in man concluded, "The general consensus . . . , based on experiments with animals and observations on man, is that DDT used with discretion does not constitute a hazard to human health. The case here recorded is the exception which tests the rule."[7]

The cynical reader aware of DDT's subsequent history will already be sighing, "*Plus ça change. . . .*" Yet despite this repetition of so much earlier history, the experience with lead arsenate and other arsenical insecticides was not entirely squandered. The years of wrestling with arsenic and lead residues had made scientists and health officials sensitive to the possible danger of small quantities of a regularly ingested chemical and to the necessity of detailed and extended laboratory studies of the chronic pathological effects of such chemicals. Samples from the earliest shipments of DDT to the United States were used by both the Public Health Service and the FDA's Division of Pharmacology for acute and chronic toxicity investigations.[8] The Public Health Service experiments lasted only a few months and were much too brief to provide any meaningful information on DDT's chronic toxicity,[b] but they never-

[b] The Public Health Service, as in its Wenatchee study on lead arsenate, tended to show little concern for low level chronic damage. In the case of DDT, this attitude may be partially understood as a rationalization encouraged by the desire to use DDT to solve other public health problems such as typhus and malaria. In the instance of the experiments cited above, haste in the evaluation of the chronic effects of DDT was necessitated by pressure from the military, desperate for an insecticide to protect American fighting men.

theless reflected an appreciation that long-term damage must be evaluated. The work conducted at the Division of Pharmacology, furthermore, was carried on for several years, through the life span of the experimental animals, and detected liver damage as a result of prolonged consumption of DDT. In the years to follow, charges that DDT was responsible for chronic illness would become too numerous for citation here.

The failure to escape a DDT-residue hazard was due less to ignorance that residues might be dangerous (which had allowed the lead arsenate hazard to develop), than to the lack of legal power to prohibit the sale and use of DDT until its safety could be determined. As a result of the sulfanilamide disaster, the Food, Drug, and Cosmetic Act included a provision requiring any manufacturer of a new drug to investigate its toxicity and report his findings to the Secretary of Agriculture, whose approval was required *before* the drug could be sold to the public. The act had no similar provisions for food products or additives, however, so that as in the days of Paris green, new insecticides could be marketed without prior testing of residue toxicity. By 1950 the number of new insecticides, as well as food additives like preservatives, bread softeners, and hormones to accelerate animal growth, had increased to such numbers as to create a national uneasiness about the wholesomeness of the food supply. A Select Committee of the House of Representatives, chaired by James Delaney of New York, was appointed in that year and charged with investigating the use of chemicals in food products and recommending legislation to alleviate any hazards uncovered. Throughout its nearly three weeks of interrogation of medical scientists, food and drug officials, and food industry representatives,[c] the Committee's attention kept re-

c The International Apple Association's Samuel Fraser was naturally in Washington for the hearings, and offered the committee members his wisdom on the question of arsenical residues. "I have shown in the case of apples that we were not to have more than so much arsenic on turning to the need for a requirement for the food industry to demonstrate the safety of its additives before putting them on sale. In 1954, the "Miller amendment" to the Food, Drug, and Cosmetic Act required the manufacturer of any new pesticide chemical to include data from toxicity and residue studies in his petition to the Secretary of Health, Education, and Welfare for permission to market the product. On the basis of this and other data, the Secretary might then establish a residue tolerance (the amendment was named for A. L. Miller of Missouri, a member of the Delaney Committee; in 1958, the "food additives amendment" extended the required pre-sale testing to food additives of all sorts).

DDT and the other new organic insecticides nevertheless remained controversial, both because of the uncertainty of even the most painstaking chronic toxicity studies, and because of their more clearly demonstrable effects on creatures other than man. The major difference between the hazards of the lead arsenate and the DDT eras, in fact, is that the later insecticides have been so much more destructive to the entire fabric of life. Ecological damage—notably the poisoning of bees—was associated with the arsenicals,[d] but DDT and related pesticides were very

them, but in dried codfish you could have 13 times as much; so if the Catholics are good and eat fish every Friday, they ought to be dead, if it is so detrimental, but they are not, they have proved to be the most virile of the population. And then we find that arsenic is essential in life . . . children probably eat apples so vigorously to get their share of arsenic essential for life." *Chemicals in Food Products. Hearings before the House Select Committee to Investigate the Use of Chemicals*, Washington, 1951, pp. 669, 683.

[d] In his pre-Paris green appeal to search for better insect poisons, federal entomologist Townend Glover had noted that if "birds should perish from feeding upon these poisoned insects, it will somewhat mitigate against the advantages of the plan" (USDA *Annual Report*, 1855, pp. 66-67). Once arsenicals came into use, entomologist D. W. Coquillet examined this risk and judged it negligible. Birds in California's San Joaquin Valley, he reported in 1891, had not been "killed in any considerable numbers from having eaten . . . locusts that had

quickly found to be dangerous to many beneficial insects, as well as to birds, fish, and other wildlife. As early as 1946, a writer in *The New Republic* described the "dynamite of DDT":[9]

"On May 23, 1945, the sun shone warmly on a large oak forest near the village of Moscow, Pennsylvania. Bird calls and songs rang through the woodland as the birds flew about feeding hungry young ones. But the forest was ill; its leaves were covered with millions of devouring gypsy moth caterpillars. Though birds ate vast numbers of the caterpillars and carried them to their newly hatched young, the horde was beyond their control.

"Early the next morning, an airplane droned over the forest, dropping a fine spray of DDT in an oil solution at the rate of five pounds an acre. The effect was instantaneous. The destructive caterpillars, caught in the deadly rain, died by the thousands. On May 25, the sun arose on a forest of great silence—the silence of total death. Not a bird call broke the ominous quiet."

In style and sentiment, these lines are identical to the book published in 1962 that raised the previously simmering discontent with DDT to a boil and made insecticidal contamination of food and the environment a subject of national alarm. The author of *Silent Spring*, Rachel Carson, was a biologist employed by the Fish and Wildlife Service who excelled at interpreting her science for the public. The near-poetic description of *The Sea Around Us* was her first bestseller, and revealed that devotion to environmental wholeness which by the late 1950s was driving her from a preoccupation with expounding the beauties of

---

been poisoned [by arsenic]." Rabbits, on the other hand, "were destroyed in large numbers," though Coquillet's way of describing this result ("at least two dozen hares [on the single plantation where he conducted his investigation] paid the penalty with their lives") suggests an indifference toward ecological disruption that was generally characteristic of the arsenical period. (U.S. Department of Agriculture, Bureau of Entomology, *Bulletin 25*, 1891, p. 60.)

biological science to one with exposing the horrors of chemical technology. The new pesticides most infuriated her, and *Silent Spring* was the product of her rage. It reported not only numerous instances of birds destroyed by carelessly applied pesticides, but also examined in taut, angry prose the damages, demonstrated and potential, that these chemicals inflicted on all forms of plant and animal life, on water and soil, and on human beings. Pesticidal poisoning, Rachel Carson concluded, had gone "beyond the dreams of the Borgias."[10]

Her chapter of that title introduced *Silent Spring*'s readers to the dangers of human poisoning from spray residues, though by simply changing the names of the new insecticides, wherever they occur, to lead arsenate, the chapter can be made to read like a summary of the years of arsenic and lead-residue regulation. The largely unrestricted use of agricultural chemicals that allowed farmers to spray too frequently, too late in the season, and with too concentrated mixtures was deplored; the difficulties of evaluating the danger of small residues and of establishing a tolerance that was both protective and enforceable were explained; the weaknesses of the Food and Drug Administration, from its inadequate funding to the limitation of its regulative power to interstate commerce, to its lack of analytical techniques applicable to some insecticides, to its exposure to opposition from the food industry were detailed; the argument for using less toxic insecticides, as well as developing non-chemical means of control, was presented. Finally, the absurdity of poisoning food to save it was suggested by the parable of Lewis Carroll's White Knight, who thought of "a plan to dye one's whiskers green, and always use so large a fan that they could not be seen."

One of the many critics of *Silent Spring* said it reminded him of *100,000,000 Guinea Pigs*, which he had read as a youth, because it was "more emotional than accurate."[11] The book was emotional, but it was also largely accurate, and it pinpointed the risk that arises when technological progress is not disciplined. Miss Carson wrote: "It is not my contention that chemical insecticides must never be used. I do contend that we have put poisonous and biologically potent chemicals indiscriminately into the hands of persons largely or wholly ignorant of their potentials for harm. We have subjected enormous numbers of people to contact with these poisons, without their consent and often without their knowledge. . . . I contend, furthermore, that we have allowed these chemicals to be used with little or no advance investigation of their effect on soil, water, wildlife, and man himself. Future generations are unlikely to condone our lack of prudent concern for the integrity of the natural world which supports all life."[12]

These remarks were intended as an introduction to *Silent Spring*'s discussion of modern pesticide abuses, but they apply equally well to the earlier unrestrained, ad hoc development of the arsenical insecticides; they are equally appropriate as a preface to our present spray-residue dilemma or as an epitaph to the pre-DDT era.

Bibliographic Notes

Preface

[1] G. Ordish, *World Review Pest Control*, 7, 204 (1968).

[2] F. Graham, *Since Silent Spring*, Boston, 1970, p. xii.

[3] *ibid.*, p. 165.

[4] P. Carles and L. Barthe, *Bull. Soc. Chim.* [4] *11*, 417 (1912).

Chapter One

[1] A great deal has been written on the history of American agriculture. The following were particularly useful for this study: P. W. Bidwell and J. I. Falconer, *History of Agriculture in the Northern United States, 1620-1860*, Washington, 1925; P. W. Gates, *The Farmer's Age: Agriculture, 1815-1860*, New York, 1960; F. A. Shannon, *The Farmer's Last Frontier. Agriculture, 1860-1897*, New York, 1945; J. Schaefer, *The Social History of American Agriculture*, New York, 1936; W. H. Clark, *Farms and Farmers: The Story of American Agriculture*, Boston, 1945. The various factors influencing the nineteenth-century industrialization of American agriculture are also discussed by G. Borgstrom, "Food and Agriculture in the Nineteenth Century," in M. Kranzberg and C. W. Pursell, Jr. (eds.), *Technology in Western Civilization*, New York, 1967, volume 1, pp. 408-424. The relation between the modernization of agriculture and the intensification of the insect problem is treated by A. J. Ihde, "Pest and Disease Controls," *ibid.*, volume 2, pp. 369-385.

[2] Quoted by Bidwell, p. 119.

[3] R. B. Morris (ed.), *Encyclopedia of American History*, New York, 1953, p. 442.

[4] Bidwell, p. 305.

[5] Cited in United States Department of Agriculture (hereafter USDA), Bureau of Entomology, *Bulletin 6*, 1896, p. 9.

[6] *First Annual Report of the U.S. Entomological Commission for 1877*, Washington, 1878, p. 95.

[7] *ibid.*, p. 82.

[8] USDA *Annual Report*, 1877, p. 266.

[9] *ibid.*, p. 265.

[10] Quoted in E. Tilton, *Amiable Autocrat*, New York, 1947, p. 207.

[11] Quoted in L. O. Howard, *A History of Applied Entomology (Somewhat Anecdotal)*, Washington, 1930, p. 202. Howard's work is the most informative and entertaining of the several histories of entomology, and has been used freely by the author in his discussion of the development of economic entomology. Also very useful is H. O. Osborn, *Fragments of Entomological History*, Columbus, Ohio, 1937, 2 volumes.

[12] Cited by editor, *American Entomologist, 1,* 3 (1868).

[13] Howard, p. 71.

[14] Osborn, p. 239.

[15] See, for example, *Practical Entomologist, 1,* 39 (1865).

[16] Howard, p. 167.

[17] *ibid.*, p. 106.

[18] USDA *Annual Report*, 1864, p. 564.

[19] *Practical Entomologist, 1,* 4 (1865).

[20] USDA *Annual Report*, 1865, p. 33.

[21] Harris, *A Treatise on Some of the Insects of New England which are Injurious to Vegetation*, Boston, 1862, 3rd edition, pp. 351, 139, and 63.

[22] *Report of the Commissioner of Patents. Agriculture*, Washington, 1854, p. 58.

[23] USDA *Annual Report*, 1871, p. 88.

[24] Quoted in B. Lehane, *The Compleat Flea*, New York, 1969, p. 29.

[25] Quoted in H. B. Weiss, *J. Econ. Ent.,* 5, 88 (1912).

[26] A more detailed history of pyrethrum may be found in USDA *Annual Report*, 1881-1882, p. 77f.

[27] *ibid.*, p. 83.

[28] C. V. Riley, *American Entomologist, 3*, 193 (1880).

[29] USDA *Annual Report*, 1862, p. 375.

[30] *Report of the Commissioner of Patents*, 1855, pp. 66-67.

[31] *Practical Entomologist, 1*, 4 (1865).

[32] B. D. Walsh, *Practical Entomologist, 1*, 1 (1865).

[33] *American Agriculturalist, 27*, 283 (1868).

[34] *ibid., 28*, 285 (1869).

[35] *Practical Entomologist, 2*, 15 (1867).

[36] E. O. Essig, *A History of Entomology*, New York, 1931, p. 50.

[37] The earliest claim for the use of Paris green seems to be that of Byron Markham, a Michigan farmer who many years later recalled dusting his potato vines with the arsenical in 1867; see *Insect Life, 5*, 44 (1892-1893).

[38] *American Agriculturalist, 27*, 321 (1868).

[39] *ibid., 30*, 207 (1871).

[40] Even the poor woodchuck was poisoned with Paris green treated cabbage leaves. *American Agriculturalist, 39*, 23 (1880).

[41] *American Agriculturalist, 37*, 183 (1878).

[42] Quoted by Osborn, volume 1, p. 150.

[43] Howard, p. 115.

[44] T. H. Haskins, *Garden and Forest, 4*, 247 (1891).

[45] T. Glover, USDA *Annual Report*, 1870, p. 75.

[46] W. McMurtrie, USDA *Annual Report*, 1874, p. 152f; 1875, p. 144f.

[47] *American Agriculturalist, 38*, 292 (1879).

[48] A. J. Cook, Michigan Agricultural Experiment Station (hereafter AES), *Bulletin 53*, 1889.

[49] *American Bee Journal, 24*, 40 (1888); 25, 355, 387, 420 (1889).

[50] *Insect Life, 4*, 282 (1891-1892); J. A. Lintner, *ibid., 6*, 181-185 (1893-1894).

[51] F. M. Webster, *Insect Life, 2*, 84-85 (1889-1890).

[52] F. M. Webster, *Insect Life, 5*, 121 (1893).

[53] *ibid., 7*, 132-134 (1894-1895).

[54] C. H. Fernald, USDA Bureau of Entomology, *Bulletin 6*, 1896, p. 8.

[55] *American Agriculturalist, 42*, 453 (1883).

[56] W. H. White, *Country Gentleman, 48*, 334-335 (1883).

[57] S. L. Allen, *ibid., 42*, 168 (1877).

[58] *American Agriculturalist, 36*, 274-275 (1877).

[59] *ibid., 37*, 315 (1878).

[60] Discussed in *American Agriculturalist, 38*, 251 (1879).

[61] A. J. Cook, *Proceedings of the Society for the Promotion of Agricultural Science, 1*, 112-114 (1880-1882). This study was still being cited by farmers and entomologists two generations later as proof of the safety of arsenicals.

[62] L. R. Taft, *Science, 21*, 259-260 (1893); R. C. Kedzie, Michigan AES, *Bulletin 101*, 1893.

[63] J. Fletcher, *Experiment Station Record, 4*, 437 (1892-1893); H. Garman, Kentucky AES, *Bulletin 53*, 1894.

[64] W. P. Headden, Colorado AES, *Bulletin 131*, 1908.

[65] USDA *Annual Report*, 1891, p. 376.

[66] For examples of poisoning reports, see *Country Gentleman, 41*, 275 (1876); *Insect Life, 1*, 142 (1888-1889).

[67] *American Agriculturalist, 36*, 274-275 (1877).

[68] C. V. Riley, *American Entomologist, 3*, 246 (1880).

[69] *ibid.*, p. 193.

[70] USDA *Annual Report*, 1887, p. 103.

[71] E. G. Lodeman, *The Spraying of Plants*, Norwood, Mass., 1896, pp. 231-232.

[72] *American Agriculturalist, 42*, 453 (1883).

Chapter Two

[1] Cited by R. Christison, *A Treatise on Poisons*, Edinburgh, 1832. Consult Christison and P. Brouardel, *Les Intoxications*, Paris, 1904, for the most instructive examples of criminal ingenuity in the use of arsenic.

[2] Discussed in *Boston Med. Surg. J., 119*, 70 (1888).

[3] Quoted by M. Morris, in T. Oliver, *Dangerous Trades*, London, 1902, p. 378.

[4] Christison, *A Treatise on Poisons*, Edinburgh, 1845, p. 223.

[5] Articles criticizing arsenical merchandise are too numerous to list *in toto*. For good general reviews, see F. C. Shattuck, *Boston Med. Surg. J., 128*, 540-546 (1893), and H. A. Lediard, *Transactions of the Sanitary Institute of Great Britain, 4*, 110-120 (1883).

[6] *Med. Surg. Reporter, 49*, 471 (1883).

[7] *Med. News*, 1883, pp. 333-334.

[8] R. Hunt, *Lancet*, 1837-1838 (i), p. 324. Also see *Lancet*, 1836-1837 (ii), pp. 556-557.

[9] *Med. Record,* 55, 354 (1899).

[10] Shattuck, p. 554.

[11] Quoted in F. H. Brown, *Boston Med. Surg. J.,* 94, 534 (1876).

[12] Reprinted in *Chemical News,* 26, 29-31, 39-41, 52-54, 90-92, 102-104 (1872).

[13] Editorial, *Boston Med. Surg. J., 114,* 233 (1886).

[14] Discussion, *Boston Med. Surg. J., 116,* 130 (1887).

[15] G. S. Hale, *Science, 19,* 104 (1892).

[16] Statement by C. F. Chandler, quoted in editorial, *Boston Med. Surg. J., 114,* 402 (1886).

[17] Quoted in Christison, *A Treatise on Poisons,* Edinburgh, 1832, p. 288.

[18] J. J. Putnam, *Boston Med. Surg. J., 124,* 623 (1891).

[19] W. B. Hills, *Boston Med. Surg. J., 131,* 453-455 (1894).

[20] Putnam, pp. 623-624.

[21] Putnam, *Boston Med. Surg. J., 119,* 1-4 (1888).

[22] Putnam, *Boston Med. Surg. J., 124,* 623 (1891).

[23] Eighteenth-century theoretical medicine is too broad a subject to be treated in any detail here. For further information on this subject, as well as on the nineteenth-century reaction against systematic medicine (including an analysis of the program and influence of the Paris school mentioned later in the text), see R. H. Shryock, *The Development of Modern Medicine,* New York, 1947, Chapters 2, 3, 4, 9, and 10.

[24] *The Scalpel, 1,* 253-254 (1849).

[25] Quoted by F. Trollope, *Domestic Manners of the Americans,* New York, 1949, p. 84.

[26] O. W. Holmes, *Medical Essays,* Boston, 1899, pp. 203, 252-253.

[27] The introduction of Fowler's solution into medicine is discussed by E. Kremers and G. Urdang, *History of Pharmacy,* Philadelphia, 1940, p. 110.

[28] I. Dyer, *Med. News, 65,* 227-230 (1894).

[29] J. Aulde, *New York Med. J., 53,* 390-397 (1891).

[30] D. G. Brinton, *Med. Surg. Reporter, 39,* 193 (1878).

[31] J. H. Winslow, *Darwin's Victorian Malady. Evidence for its Medically Induced Origin,* Philadelphia, 1971.

[32] This growing anti-arsenicism was a natural reaction against the profession's extreme reliance on Fowler's solution and was also encouraged by increasing appreciation of the possibility of chronic arsenic poisoning. For examples of criticism of arsenical

therapy see *J. Cut. Ven. Dis.*, *4*, 179, 218-220, 362-365, 366-377 (1886); J. J. Putnam, *Boston Med. Surg. J.*, *119*, 1-4 (1888); Dyer, *Med. News*, *65*, 227 (1894).

[33] P. Pott, *The Chirurgical Observations of Percival Pott*, Philadelphia, 1819, Volume 2, pp. 291-295.

[34] J. A. Paris, *Pharmacologia*, New York, 1822, pp. 208-209. The original edition was published in London, 1820.

[35] H. T. Butlin, *Brit. Med. J.*, 1892 (i), 1,341-1,346; (ii), 1-6, 66-71.

[36] For pre-Hutchinson reports of skin cancer following arsenical medication of psoriasis and other skin ailments, see M. Pozzi, *Bull. Soc. Anat. Paris*, 5 (3^e), 587-588 (1874); A. Cartaz, *ibid.*, 2, (4^e), 549-550 (1877); J. C. White, *Am. J. Med. Sci.*, *89*, 163-173 (1885); H. Hebra, *Monats. f. prakt. Derm.*, *6*, 1-9 (1887). Hutchinson's article, "Arsenic Cancer," appeared in *Brit. Med. J.*, 1887 (ii), 1,080-1,081. For a discussion of the relation of arsenicals to cancer after Hutchinson, see O. Neubauer, *Brit. J. Cancer*, *1*, 192-251 (1947).

[37] For an English rendering of von Tschudi's accounts, see *Boston Med. Surg. J.*, *51*, 189-195 (1855).

[38] J. Johnston, *Chemistry of Common Life*, New York, 1855, Volume 2, p. 166.

[39] S. O. L. Potter, *A Handbook of Materia Medica, Pharmacy and Therapeutics*, Philadelphia, 1903, p. 199. Also, C. Maclagan, *Boston Med. Surg. J.*, *71*, 200 (1864-1865).

[40] E. W. Schwartze, *J. Pharmacol. Exp. Therap.*, *20*, 181-203 (1922). Even Schwartze's work failed to bring about complete disavowal of the Styrian phenomenon. Subsequent toxicological writers continued to recognize the possibility of developing some degree of tolerance to the poison. See T. Sollman, *A Manual of Pharmacology*, Philadelphia, 1948, p. 873; L. Goodman and A. Gilman, *The Pharmacological Basis of Therapeutics*, New York, 1941, p. 740; C. Thienes and T. Haley, *Clinical Toxicology*, Philadelphia, 1948, p. 170. Considerably less reserve has been demonstrated by others. In M. Marten and N. Cross's *The Doctor Looks at Murder*, New York, 1937, extracted in *Literary Digest*, *124*, 24 (1937), the authors state that "the girls in Styria are more beautiful than others anywhere in Europe. In particular, arsenic produces clear, lovely complexions and hair of unsurpassed glowing warmth and beauty."

[41] Christison, 1845 edition, p. 34.

[42] Statement by a Dr. Kesteven, quoted by Maclagan, p. 204.

[43] *Boston Med. Surg. J., 51*, 190 (1855).

[44] The Boston neurologist Putnam, for example, asserted that Styrians sometimes were paralyzed by chronic arsenical intoxication: *Boston Med. Surg. J., 119*, 1-4 (1888).

[45] D. M. Parker, *Edinburgh Med. J., 10*, 116-123 (1864).

[46] Quoted, *Boston Med. Surg. J., 39*, 425 (1849).

[47] Shattuck, p. 543. This address was delivered before the Philadelphia Pathological Society.

[48] *Med. Record, 39*, 599-600 (1891).

[49] *ibid., 55*, 360 (1899).

[50] W. B. Hills, *Boston Med. Surg. J., 131*, 454 (1894).

[51] R. C. Kedzie, Michigan AES, *Bulletin 101* (1893), p. 21.

[52] Quoted in S. Smith, *The City That Was*, New York, 1911, pp. 66-67.

[53] H. I. Bowditch, *Public Hygiene in America*, Boston, 1877, p. 29.

[54] *ibid.*, p. 122.

[55] Lediard, p. 119.

[56] Bowditch, p. 51.

[57] For stories relating to the grape scare, see *New York Times*, September 25 to September 27, 1891.

[58] B. T. Galloway, USDA *Annual Report*, 1891, p. 376.

Chapter Three

[1] See *Brit. Med. J.*, 1892 (i), p. 741, for a discussion of the British objections to American fruit.

[2] W. M. Munson, Maine AES, *Annual Report*, 1891, pp. 81-122.

[3] S. T. Maynard, Massachusetts AES, *Bulletin 17*, 39 (1892).

[4] Quoted in A. Gautier, *Bull. Acad. Med.*, 70, 375 (1913).

[5] *American Association of Economic Entomologists, Presidential Addresses*, mimeographed by USDA, Bureau of Entomology and Plant Quarantine, Washington, 1889-1911. See 1898 address by H. Osborn, p. 2.

[6] F. W. Sempers, *Injurious Insects and the Use of Insecticides*, Philadelphia, 1893, p. 52.

[7] For examples, see S. A. Beach and L. L. Van Slyke, New York AES, *Bulletin 41*, 1892; Maynard; W. B. Alwood, Virginia

AES, *Bulletin 15*, 1892; C. M. Weed, *Popular Science Monthly*, *42*, 638-647 (1892-1893); USDA *Farmer's Bulletin 7*, 1892.

[8] Maynard, p. 39.

[9] Beach and Van Slyke, p. 58.

[10] USDA *Farmer's Bulletin 7*, p. 10.

[11] See E. G. Lodeman, *The Spraying of Plants*, Norwood, Mass., 1896, Appendix A, for a discussion of spraying legislation.

[12] F. C. Sears, *Productive Orcharding*, Philadelphia, 1923, 3rd edition, p. 201. The book's first edition was in 1914.

[13] A. L. Melander, *J. Econ. Ent.*, *1*, 219 (1908).

[14] H. Garman, Kentucky AES, *Bulletin 53*, 125-143 (1894).

[15] H. D. Gibbs and C. C. James, *J. Am. Chem. Soc.*, 27, 1,484-1,496 (1905).

[16] S. A. Forbes, Illinois AES, *Bulletin 108*, 1906.

[17] *ibid.*, p. 281.

[18] C. D. Woods, Maine AES, *Bulletin 224*, 1914, p. 48.

[19] W. C. O'Kane, C. H. Hadley, and W. A. Osgood, New Hampshire AES, *Bulletin 183*, 1917, p. 36. O'Kane discussed his work in progress at professional entomological meetings: see *J. Econ. Ent.*, 8, 191 (1915); 9, 90-91 (1916).

[20] *ibid.*, p. 14.

[21] *ibid.*, p. 13.

[22] *ibid.*, p. 35.

[23] *ibid.*, p. 34.

[24] *ibid.*, p. 36.

[25] W. Beaumont, *Experiments and Observations on the Gastric Juice and the Physiology of Digestion*, Plattsburgh, N.Y., 1833.

[26] A. J. Carlson, *American Journal of Physiology*, *31*, 151 (1912).

[27] O'Kane, p. 28.

[28] *ibid.*, p. 27.

[29] *ibid.*, pp. 27-28.

[30] *ibid.*, p. 27.

[31] *ibid.*, p. 35. O'Kane's work also included extensive studies of the effects of arsenical sprays on livestock and poultry that indicated no danger to farm animals so long as ordinary precautions were followed during application of the spray.

[32] Quoted in Gautier, p. 375.

[33] E. S. Reynolds, *Brit. Med. J.*, 1900 (ii), 1,492-1,493.

[34] T. N. Kelynack, et al., *Lancet*, *78* (ii), 1,600 (1900).

[35] H. G. Brooke and Leslie Roberts, *Brit. J. Dermatology*, *13*, 125 (1901).

[36] S. Delepine and C. H. Tattersall, *Brit. Med. J.*, 1900 (ii), 1,587-1,588. Manchester beer was found to contain concentrations of 1/40 to 1/16 grain of arsenic per gallon.

[37] R. W. MacKenna, *Brit. Med. J.*, 1901 (i), 85.

[38] Great Britain, *Royal Commission on Arsenical Poisoning. Final Report*, London, 1903, p. iii. Many feared that this charge was too broad, that the commission would need several years to determine the dangers of arsenic in all food and drink, and that during this protracted period, unless immediate action were taken by Parliament, the beer epidemic might continue or arise in other localities (*Lancet, 79* [i], 192-193, 271-272, 414 [1901]). The epidemic was already on the wane, however, thanks largely to the self-policing efforts of the Manchester Brewer's Association. This group of merchants voluntarily withdrew from the market supplies of beer that might contain arsenic, and instituted thorough chemical testing of all further supplies before placing them on sale (Delepine). The selfless action of the English brewers should have provided both a moral and a political example for American apple growers of the 1920s.

[39] E. A. Ormerod, *Eleanor Ormerod, LL.D. Economic Entomologist. Autobiography and Correspondence*, New York, 1904, pp. 206-208. Shortly after, Dr. Ormerod found herself complaining that her countrymen had gone from one extreme to the other, and were using arsenicals recklessly (*Insect Life, 4*, 36-39 [1891-1892]).

[40] Great Britain, *Royal Commission on Arsenical Poisoning. First Report*, London, 1901, p. 90.

[41] *ibid.*, p. 280.

[42] H. Martin, *Residue Reviews, 4*, 17-32 (1963), points out that "although without statutory authority during the period from 1903 to 1960, the arsenic tolerance proposed by the Commission was invariably accepted as expert evidence in British Courts of Law. . . . The recommendation proved adequate for the control of the importation into Great Britain of unwashed apples from countries where heavy amounts of lead arsenate had been applied against codling moth. Indeed it is difficult to find any objective reason why it was considered necessary, in 1960, to give the tolerance statutory authority" (p. 26).

265

[43] J. K. Haywood, USDA Bureau of Chemistry, *Bulletin 86*, 1904, p. 25.

[44] A. Riche, *Bull. Acad. Med.*, *59*, 192-209 (1908).

[45] P. Cazeneuve, *ibid.*, pp. 134-154.

[46] G. Weiss, *Bull. Acad. Med.*, *61*, 141 (1909).

[47] G. Linnossier, *ibid.*, 61-67 (1909).

[48] A. Gautier, *ibid.*, p. 103.

[49] J. Lucas-Championniere, *ibid.*, 99-115 (1909). In Germany, too, there was opposition to the use of arsenicals. The chemist von der Heide found in 1906 that grapes from German vines sprayed with lead arsenate contained as much as 0.2 mg. of arsenic per 100 g. (about 0.014 grain of arsenic per pound, or 40 percent above the "world tolerance." See *Chemical Abstracts*, *3*, 2,838⁹ [1909]). This discovery prompted the German Imperial Health Commission to oppose the use of lead arsenate and, in fact, the compound was eventually prohibited (temporarily) from agricultural application in Germany. Other arsenical insecticides were apparently permitted, but only under close supervision, as was generally the case in other European countries (spraying legislation in Europe is discussed in *Bull. Acad. Med.*, *70*, 415-440 [1913], and *71*, 300-311 [1914]).

[50] A. Gautier, *Bull. Acad. Med.*, *70*, 371 (1913). Five years before, Cazeneuve had declared that, "The use in America of this dangerous agent for exported fruits which the Americans do not eat (!) does not suffice to persuade me [of the safety of arsenicals]" (*Bull. Acad. Med.*, *59*, 153 [1908]).

[51] E. G. Packard, *Entomological News*, *17*, 256 (1906).

Chapter Four

This and the following chapters include numerous references to unpublished government documents. To save space, a system of abbreviated citations to these documents has been adopted. Items from the Records of the Office of the Secretary of Agriculture, General Correspondence, Sprays File, Record Group 16, kept in the National Archives in Washington, D.C., have been abbreviated as Rec. Off. Sec. Ag. Material from the Records of the Bureau of Foreign and Domestic Commerce, File 331.0, United Kingdom, Record Group 151, also in the National Archives, is referred to by Rec. BFDC. Finally, the Records of the Food and Drug Administration, General Spray Residue File, Record Group

88, housed in the National Records Center, Suitland, Maryland, is abbreviated Rec. FDA. Documents in each Record Group are filed in chronological order.

[1] For a fuller discussion, see F. A. Filby, *A History of Food Adulteration and Analysis*, London, 1934.

[2] Born in Germany, Accum spent most of his professional career in London. Hassall demonstrated the usefulness of the microscope in detecting food adulterants, and employed it with great success as head of the Analytical Sanitary Commission, a food examination project commissioned by the medical journal, *The Lancet*. See his *Food and its Adulterations*, London, 1855. Also see H. A. Schuette, "Death in the Pot," *Trans. Wisconsin Acad. Sci.*, *35*, 283-303 (1943).

[3] The problem of food adulteration and pure food legislation in the United States has received a great deal of attention. Discussion of this subject in the text is particularly indebted to O. E. Anderson, Jr., *The Health of a Nation: Harvey W. Wiley and the Fight for Pure Food*, Chicago, 1958; W. H. Wiley, *An Autobiography* (originally published in 1930; I have relied on a paperback version from Rodale Books, 1957); Wiley, *The History of a Crime Against the Food Law*, Washington, D.C., 1929; G. A. Weber, *The Food, Drug and Insecticide Administration; Its History, Activities and Organization*, Baltimore, 1928; C. C. Regier, "The Struggle for Federal Food and Drugs Legislation," *Law and Contemporary Problems*, *1*, 3-15 (1933); J. H. Young, "Social History of American Drug Legislation," in P. Talalay (ed.), *Drugs in our Society*, Baltimore, 1964, pp. 217-229.

[4] Quoted in Regier, p. 12.

[5] Wiley, *An Autobiography*, p. 202.

[6] *Report of the Commissioner of Agriculture*, 1884, pp. 53-69.

[7] See Wiley, *History of a Crime Against the Food Law*, pp. 86 f., 96 f.

[8] Wiley, "Influence of Food Preservatives and Artificial Colors on Digestion and Health," USDA, Bureau of Chemistry, *Bulletin 84*, 1904, p. 8.

[9] Quoted by J. H. Young, "The Science and Morals of Metabolism: Catsup and Benzoate of Soda," *J. Hist. Med.*, *23*, 92 (1968). A fascinating article providing insight into Wiley's moral commitment to the pure food crusade, I have relied heavily on its discussion of Wiley's study of food preservatives.

[10] Detailed in Wiley, *Bulletin 84*.

[11] Written by S. W. Gillian and quoted in Wiley, *An Autobiography*, p. 217.

[12] *ibid.*, p. 217.

[13] Wiley, *History of a Crime Against the Food Law*, pp. 162-163.

[14] See Young, "The Science and Morals of Metabolism," for a discussion of the Referee Board's study of sodium benzoate and of the conflict between their findings and Wiley's Poison Squad investigations.

[15] Wiley, *Bulletin 84*, p. 1,494.

[16] Wiley, *History of a Crime Against the Food Law*, p. 24.

[17] *ibid.*, pp. 25, 10.

[18] *ibid.*, p. 41.

[19] Wiley, *Foods and Their Adulteration*, Philadelphia, 1907, pp. 334-335.

[20] See Wiley, *Bulletin 84*, pp. 754-756; *History of a Crime Against the Food Law*, pp. 50-51.

[21] Anderson, p. 207. Wiley preferred exact tolerances and had recommended a copper tolerance of 11 mg. per 100 g. of vegetable. This tolerance was relaxed by his more liberal colleagues while Wiley was abroad.

[22] Quoted *ibid.*, p. 208.

[23] W. R. M. Wharton, *Food Drug Cosmetic Law Q.*, *1*, 354 (1946); J. H. Young, *The Medical Messiahs*, Princeton, 1969, p. 45.

[24] Quoted in Anderson, p. 185.

[25] *ibid.*, p. 200.

[26] Quoted by Wiley, *History of a Crime Against the Food Law*, pp. 375-376.

[27] *ibid.*, p. 376.

[28] F. B. Linton, *Food Drug Cosmetic Law J.*, *5*, 111 (1950).

[29] *ibid.*, pp. 326-339, has an excellent discussion of Campbell's early career in the Bureau of Chemistry.

[30] Wiley, *Foods and Their Adulteration*, p. 330.

[31] Lynch, W. D., et al., "Poisonous Metals on Sprayed Fruits and Vegetables," USDA, *Bulletin 1027*, 1922.

[32] The conference minutes are contained in Rec. FDA.

[33] For a discussion of the history of the debate over "normal physiological" arsenic, see I. Bang, *Biochemische Zeitschrift*, *165*,

378-380 (1925); and W. F. Boos and A. B. Werby, *New England J. Med.*, *213*, 520-527 (1935).

[34] Letter from F. B. Linton to Alsberg, July 10, 1920. Rec. FDA.

[35] Anonymous address to New Jersey Horticultural Society, by member of the Bureau of Chemistry, p. 3. Rec. FDA.

[36] Press release, Office of Information, USDA. Rec. FDA.

[37] Letter from W. G. Campbell to Congressman E. F. Taylor, Aug. 17, 1922. Rec. FDA.

[38] *ibid.*

[39] Letter from Alsberg to Woodward, May 14, 1921. Rec. FDA.

[40] Letter from Campbell to A. L. Quaintance, Bureau of Entomology, July 30, 1921. Rec. FDA.

[41] Letter from P. B. Dunbar, Bureau of Chemistry, to Senator W. H. Borah, April 29, 1921. Rec. FDA.

[42] Anonymous address, "Regulatory Work of the U.S. Department of Agriculture," p. 4. Rec. FDA.

[43] Letter from B. C. Winslow, Los Angeles Station, to R. W. Hilts, Chief, Western District, Dec. 5, 1921. Rec. FDA.

[44] Letter from Campbell to A. L. Melander, Oct. 26, 1926. Rec. FDA.

[45] R. Lord, *The Wallaces of Iowa*, Boston, 1947, p. 295. For a detailed analysis of the factors responsible for the agricultural depression of the 1920s, see E. G. Nourse, *American Agriculture and the European Market*, New York, 1924.

[46] The arsenical celery episode is discussed in an anonymous address to the New Jersey Horticultural Society, pp. 4-6; and in R. Lamb, *American Chamber of Horrors*, New York, 1936, p. 200 f.

[47] Documents relating to the New Jersey apple seizure are in the file, "Cooperation-New Jersey, 1925-30," Rec. FDA.

[48] Anonymous address, p. 7.

[49] Letter from Wharton to H. B. Costill, Director, New Jersey State Department of Health, October 21, 1925. Rec. FDA.

[50] Anonymous address, p. 9.

[51] *ibid.*, p. 9. See also USDA, Bureau of Chemistry, *Service and Regulatory Announcements*, 1926, Notice of Judgment 13936, p. 485.

[52] Anonymous address, p. 11.

Chapter Five

[1] Speech by R. G. Phillips, Salt Lake City Spray Conference, Feb. 21, 1927. Rec. FDA. Also see H. E. Cox, *Analyst, 51,* 132-137 (1926).

[2] Speech by Phillips, p. 8.

[3] Statement by P. B. Dunbar, Conference on Arsenical and Lead Spray Residues, Philadelphia, Dec. 29, 1926, p. 2. Rec. FDA. The scientific press was only slightly more charitable. The editor of *The Lancet*, for example, submitted that "to depend upon wind and rain for the removal from the surface of apples of the arsenical compound is surely lighthearted." *Lancet, 102* (ii), 1,295 (1925).

[4] Confidential speech by R. G. Phillips, "The Spray Residue Situation," p. 8. Rec. FDA. Also see *Gardener's Chronicle, 79* (*3rd*), 38 (1926); letter from W. G. Jamison, Dept. of Commerce, to Portland district office, Feb. 13, 1926. Rec. BFDC.

[5] Quoted in *Gard. Chron., 79* (*3rd*), 427 (1926).

[6] Confidential speech by Phillips, pp. 10-11.

[7] Letter from R. G. Phillips to W. G. Jamison, July 29, 1926. Rec. BFDC.

[8] Letter from H. F. Senftner to USDA, Dec. 18, 1925. Rec. FDA.

[9] Phillips, "Arsenic Spray Residue," confidential report, p. 6. Rec. Off. Sec. Ag.

[10] Comment by P. B. Dunbar at Arsenical and Lead Spray Residues Conference, Philadelphia, Dec. 29, 1926. Rec. FDA.

[11] Speech by W. G. Campbell, Salt Lake City Spray Conference, Feb. 21, 1927. Rec. FDA.

[12] Letter from C. L. Alsberg to W. C. Woodward, May 14, 1921. Rec. FDA.

[13] Memo, May 10, 1921. Rec. FDA.

[14] Memorandum from C. A. Browne to W. Jardine, Secretary of Agriculture, Dec. 4, 1926 (Rec. FDA) cites 0.03 grain per pound as the Bureau's "present tolerance." There are implications in this memo and other Bureau records that the tolerance had been a bit higher in previous years.

[15] Minutes of Conference Relative to Arsenic Spray Residues on Fruits, April 27, 1926, p. 4. Rec. Off. Sec. Ag.

[16] Speech by Campbell, p. 4.

[17] Letter from E. R. Kelly to W. Jardine, Dec. 17, 1926. Rec. FDA.

[18] Letter from C. A. Browne to various toxicologists, Aug. 26, 1926. Rec. FDA.

[19] Letter from W. Wharton to P. B. Dunbar, Sept. 23, 1926. Rec. FDA.

[20] The Banks trial is discussed in some detail by Ruth Lamb, *American Chamber of Horrors*, New York, 1936, p. 211f. Also see letter from C. A. Browne to Grasselli Chemical Co., Dec. 18, 1926. Rec. FDA; *Notices of Judgment*, 15350.

[21] *ibid.*

[22] *Notices of Judgment*, 17274, pp. 173-174.

[23] Trial excerpts are taken from enclosure, letter from L. D. Elliott to G. S. Luckett, Chief, Division of Sanitary Engineering and Sanitation, Santa Fe, N.M., May 23, 1927. Rec. FDA.

[24] *Notices of Judgment*, 17274, pp. 174-175.

[25] Lamb, p. 223.

[26] *Notices of Judgment*, 17274, p. 176.

[27] Memo, Dec. 4, 1926. Rec. FDA.

[28] The minutes of the morning session of the Hunt Conference and the report of the afternoon meeting are filed in Rec. FDA.

[29] Memo from C. A. Browne to A. F. Woods, Director of Scientific work, Bureau of Chemistry, Dec. 23, 1926. Rec. FDA.

[30] Among the more pertinent studies of arsenic toxicology available to the Hunt Committee: J. K. Haywood, USDA *Bur. Chem., Bulletin 113*, 1908; W. D. Harkins and R. E. Swain, *J. Am. Chem. Soc., 30,* 928-946 (1908); A. J. Carlson, *Am. J. Physiol., 31,* 151-168 (1912); S. Ayres, *Arch. Derm. Syph., 2,* 747-756 (1920); T. Sollmann, *J. Pharmacol. Exp. Therap., 18,* 43-49 (1921); B. Throne, et al., *N.Y. State Med. J., 26,* 843-848 (1926).

[31] An exceptionally thorough treatment of the history of lead poisoning is L. G. Stevenson, *A History of Lead Poisoning,* unpub. Ph.D. dissertation, Johns Hopkins, 1949.

[32] J. C. Aub, et al., *Medicine, 4,* 186 (1925).

[33] In addition to *Public Health Bulletin, 126,* see D. Hunter, *The Diseases of Occupations,* Boston, 1969, p. 283.

[34] T. Sollmann, *J. Pharmacol. Exp. Therap., 19,* 375-384 (1922).

[35] C. Howard, *Am. J. Pub. Health, 13,* 207-209 (1923).
[36] Campbell's Salt Lake City speech is filed in Rec. FDA, p. 5.
[37] *ibid.,* p. 7.
[38] *ibid.,* p. 6.
[39] *ibid.,* p. 6.
[40] *ibid.,* pp. 7-8.
[41] *ibid.,* p. 11.
[42] Speech by Phillips, pp. 22, 25.
[43] Letter from L. D. Elliott to P. L. Smithers, Canon City, Colo., July 18, 1927. Rec. FDA.
[44] Letter from G. J. Morton, Chief, San Francisco Station, to J. W. Hebert, Chief, Western District, March 5, 1928. Rec. FDA.
[45] A good discussion of residue removal techniques is to be found in H. C. Diehl, et al., USDA *Circular 59,* 1929. A full discussion of the government's role in the residue removal project is provided by P. Dunbar, *Food Drug Cosmetic Law J., 14,* 129 (1959).
[46] H. E. Fitch, *Yakima Valley Progress,* October 1923.
[47] *The Wenoka Apple Book,* publication of the Wenatchee-Okanogan Cooperative Federation; publication date not given.
[48] Letter to M. L. Dean, Sept. 9, 1926. Rec. FDA.
[49] Letter to Sen. W. C. Jones, Sept. 10, 1926. Rec. FDA.
[50] Letter to W. M. Jardine, Dec. 23, 1927. Rec. FDA.
[51] Confidential letter from R. G. Phillips to W. G. Campbell, Aug. 6, 1927. Rec. FDA.
[52] Letter from W. G. Campbell to R. G. Phillips, Aug. 9, 1927. Rec. FDA.
[53] Letter to W. Jardine, Nov. 15, 1926. Rec. FDA.
[54] Market Letter No. 5, The Associated Fruit Company, Delta, Colorado, Sept. 29, 1927. Rec. FDA.
[55] Letter to W. Jardine, Dec. 18, 1926. Rec. FDA.
[56] Letter to Sen. L. C. Phipps, Oct. 11, 1927. Rec. FDA.
[57] *Big "Y" Bulletin,* Publication of Yakima Fruit Growers' Association, March 1934.
[58] *ibid.,* February 1927.
[59] *ibid.,* January 1927.
[60] Letter to G. J. Morton, Chief, San Francisco Station, March 7, 1928. Rec. FDA.
[61] Questionnaire from W. G. Campbell, Feb. 24, 1928. Rec. FDA.

[62] A table of arsenic and lead tolerances to 1940 may be found in H. H. Shepard, *The Chemistry and Action of Insecticides*, New York, 1951, p. 403.

[63] *Congressional Record, 69*, 6,166 (1928).

[64] H. Wiley, *Good Housekeeping*, July 1928, p. 100.

[65] *Congressional Record, 69*, 9,524-9,531 (1928).

[66] E. R. Smith, *Amer. Pomol. Soc., Proc., 46*, 55 (1929).

Chapter Six

[1] *Science, 67*, 357 (1928).

[2] *Am. J. Pub. Health, 19*, 1,052-1,053 (1929).

[3] B. Throne, et al., *N.Y. State J. Med., 26*, 843-848 (1926).

[4] C. N. Myers and B. Throne, *N.Y. State J. Med., 29*, 1,259 (1929).

[5] C. N. Myers, et al., *Ind. Eng. Chem., 25*, 624 (1933).

[6] Myers, note 4 above.

[7] L. S. Van Dyck, B. Throne, and C. N. Myers, *Arch. Ped., 47*, 218-229 (1930).

[8] A. F. Kraetzer, *J. Am. Med. Assoc., 94* (ii), 1,036 (1930).

[9] W. D. Shelden, et al., *Arch. Neurol. Psych., 27*, 322-332 (1932).

[10] C. N. Myers, et al., *Ind. Eng. Chem., 25*, 624-628 (1933).

[11] S. Ayres and N. P. Anderson, *Arch. Derm. Syph., 30*, 35 (1934).

[12] C. N. Myers, et al., *N.Y. State J. Med., 35*, 579-589 (1935).

[13] *J. Am. Med. Assoc., 104* (i), 220 (1935).

[14] *Am. J. Pub. Health, 26*, 388 (1936).

[15] A. B. Cannon, *N.Y. State J. Med., 36*, 233 (1936).

[16] P. J. Hanzlik, *Sci. Mon., 44*, 436 (1937).

[17] M. McNicholas, *A Study of Some Effects of Ingested Arsenious Oxide*, Washington, D.C., 1937, p. 33.

[18] *J. Am. Med. Assoc., 99*, 2,202-2,203 (1932).

[19] W. F. Cogswell and J. W. Forbes, *Am. J. Pub. Health, 26*, 380 (1936).

[20] Quoted in Cannon, p. 232.

[21] T. J. Talbert, *Proc. Amer. Soc. Hort. Sci., 32*, 170-174 (1934); and W. L. Tayloe, *ibid., 27*, 538-542 (1930); Missouri AES, *Research Bulletin 183*, 1933.

[22] Talbert and Tayloe, *Research Bulletin 183*, p. 4.

[23] Talbert, p. 170.

[24] A. C. Chapman, *Analyst*, *51*, 548-563 (1926); A. D. Holmes and Roe Remington, *Ind. Eng. Chem.*, *26*, 573-574 (1934); T. S. Harding, *Sci. Amer.*, *149*, 197-199 (1933).

[25] E. J. Coulson, et al., *Science*, *80*, 230-231 (1934); *J. Nutrition*, *10*, 255-270 (1935).

[26] Talbert, p. 174.

[27] These reports are tabulated and discussed in a memorandum from H. D. Lightbody to P. Dunbar, June 23, 1933. Rec. FDA.

[28] The FDA's opinion of Talbert's work is given in a memorandum from Erwin Nelson to P. Dunbar, Aug. 17, 1935. Rec. FDA.

[29] Memorandum from B. G. Winslow, Aug. 14, 1928. Rec. FDA.

[30] Letter to USDA, Feb. 18, 1929. Rec. FDA.

[31] Letter to W. M. Jardine, June 23, 1927. Rec. FDA.

[32] For a thorough discussion of muckraking journalism see: L. Filler, *Crusaders for American Liberalism*, New York, 1939; C. C. Regier, *The Era of the Muckrakers*, Chapel Hill, 1932; A. Weinberg and L. Weinberg, *The Muckrakers*, New York, 1961.

[33] J. H. Young's chapter, "The New Muckrakers," in his *The Medical Messiahs*, Princeton, 1967, offers an informative and entertaining discussion of the changes in advertising practices during the 1920s, and of the muckraking that these changes provoked.

[34] F. J. Schlink, *Introduction to Consumers' Research*, Washington, N.J., 1933.

[35] E. P. Herring, *Current History*, *40*, 33 (1934).

[36] *Your Money's Worth*, New York, 1927, pp. 256, 264.

[37] Schlink, p. 3.

[38] C. O. Jackson, *Food and Drug Legislation in the New Deal*, Princeton, 1970, p. 20.

[39] F. J. Schlink, *Con. Res. Bull.*, *3* (2), 24 (1934). Also see *ibid.*, *2* (4), 6-8 (1933); *3* (2), 1-3 (1934); *3* (3), 7 (1934); *3* (4), 23 (1934); *2* (n.s., 3), 23 (1935).

[40] *ibid.*, *2* (1), 8 (1932).

[41] *ibid.*, *3* (2), 6-7 (1934); *3* (4), 4 (1934); and R. Joyce, *ibid.*, *1* (n.s., 1), 17-21 (1934); Kallett, *ibid.*, *1* (n.s., 6), 6 (1935).

[42] Kallett's relations with Schlink, originally close, eventually became strained and in the mid-1930s he became the leader of a splinter group that formed Consumers' Union, a competitive consumer organization.

[43] *100,000,000 Guinea Pigs*, New York, 1933, p. 6.

[44] *ibid.*, p. 97.

[45] *ibid.*, p. 47.

[46] *ibid.*, pp. 59-60.

[47] *ibid.*, pp. 302-303.

[48] Letter to Secretary of Agriculture, Jan. 8, 1934. Rec. Off. Sec. Ag.

[49] Letter to Mrs. Franklin Roosevelt, Dec. 26, 1933. Rec. Off. Sec. Ag.

[50] *Eat, Drink and Be Wary*, Washington, N.J., 1935, p. 31.

[51] *40,000,000 Guinea Pig Children*, New York, 1937, p. 141.

[52] Kallett and Schlink, p. 52.

[53] *ibid.*, pp. 52, 291.

[54] Both comments quoted by Jackson, p. 19.

[55] The inadequacies of the old law and the growth of sentiment to change it are discussed in R. Lamb, *American Chamber of Horrors*, New York, 1936, p. 278f.; by Young, *The Medical Messiahs*, pp. 36f., 47f., 158f.; and by Jackson, p. 3f.

[56] Cited by Lamb, pp. 15, 27.

[57] *Congressional Record*, 62 Congress, 1 session, 2,379-2,380.

[58] Quoted by Young, p. 157.

[59] Jackson, pp. 209-210, and Young, "Social History of American Drug Legislation," p. 226, discuss the influences that discouraged popular support of food and drug reform in the 1930s.

[60] The following episode is recalled by Tugwell in *FDA Papers*, June 1968, pp. 4-8; also by Lamb, p. 278f., and Jackson, pp. 3, 24f. A thorough analysis of Tugwell's career is provided by B. Sternsher, *Rexford Tugwell and the New Deal*, New Brunswick, N.J., 1964.

[61] Quoted by Lamb, pp. 278-279.

[62] *ibid.*, p. 279.

[63] P. Dunbar, *Food, Drug, Cosmetic Law J.*, *14*, 134 (1959).

[64] Jackson gives an excellent recent analysis of the legislative history of the 1938 act. Earlier works include C. W. Dunn, *Federal Food, Drug, and Cosmetic Act: A Statement of its Legislative Record*, New York, 1938; H. A. Toulmin, Jr., *A Treatise on the Law of Food, Drugs and Cosmetics*, Cincinnati, 1942; a very informative article by D. F. Cavers is in *Law and Contemporary Problems*, *6*, 2-42 (1939).

[65] The script of the radio broadcast is filed in Rec. FDA.

[66] Letter from Dunbar to Sen. R. S. Copeland, March 13, 1935. Rec. FDA.

[67] Letter from L. A. Strong to M. Eisenhower, April 27, 1935. Rec. FDA.

[68] Lamb, p. 4.

[69] *ibid.*, p. 237.

[70] Jackson, p. 166.

[71] Letter to F. D. Roosevelt, Nov. 18, 1937. Rec. FDA.

[72] Report, "Method for Determination of Arsenic and Lead on Fruits and Vegetables," by H. J. Wichmann and C. W. Murray, May 27, 1933. Rec. FDA. The new method, which also gave a quick analysis for arsenic (arsenic had previously been determined by the Gutzeit method), consisted of digesting the original sample with acid, distilling arsenic off as the trichloride, and then titrating arsenic with bromate. After adjustment of the pH of the residue to 2.8-3.5, lead was precipitated as the sulfide, filtered, redissolved, then precipitated as chromate. The chromate was then redissolved and titrated with iodine.

[73] Letter to F. D. Roosevelt, Nov. 18, 1937. Rec. FDA.

[74] A. M. Peter, Kentucky AES, *Sixth Annual Report*, 1893, pp. 14-15; H. Garman, Kentucky AES, *Bulletin 53*, 1894.

[75] R. E. Remington, *J. Am. Chem. Soc.*, *49*, 1,410 (1927).

[76] H. Popp, *Z. Angew. Chem.*, *41*, 838-839 (1928); C. R. Gross and O. A. Nelson, *Am. J. Pub. Health*, *24*, 36-42 (1934).

[77] Remington; Gross and Nelson; C. N. Myers and B. Throne, *N.Y. State J. Med.*, *29*, 871-874 (1929).

[78] Resolution of Washington State Horticultural Association, Dec. 14, 1937. Rec. FDA.

[79] Letter from W. Campbell to J. C. Snyder, Secretary, Washington State Horticultural Association, Dec. 21, 1937. Rec. FDA.

[80] Following quotes are from L. Riggs, *Russet Mantle*, New York, 1936.

[81] Letter from W. J. Robinson to H. A. Wallace, Secretary of Agriculture, Feb. 28, 1936. Rec. Off. Sec. Ag.

Chapter Seven

[1] *Ind. Eng. Chem.*, *19*, 190 (1927).

[2] S. Marcovitch and W. W. Stanley, *J. Econ. Ent.*, *23*, 370 (1930).

[3] T. Sollmann, et al., *J. Pharmacol. Exp. Therap.*, *17*, 192-225 (1920).

[4] H. F. Smyth and H. F. Smyth, Jr., *Ind. Eng. Chem.*, *24*, 229-232 (1932).

[5] Memo of W. B. White, March 21, 1933. Rec. FDA.

[6] For further discussion of mottled enamel and its relation to dissolved fluorides, see F. J. McClure, *Water Fluoridation: The Search and the Victory*, Bethesda, Maryland, 1970.

[7] M. C. Smith and R. M. Leverton, *Ind. Eng. Chem.*, *26*, 797 (1934). Also see M. C. Smith, et al., Arizona AES, *Technical Bulletin 32*, 1931; H. V. Smith and M. C. Smith, Arizona AES, *Technical Bulletin 43*, 1932.

[8] Letter from Henry Knight to W. Campbell, June 19, 1933. Rec. Off. Sec. Ag.

[9] Summary of the Work and Findings of Floyd DeEds, Rec. Off. Sec. Ag.

[10] "Notice, etc.," March 1, 1933. Rec. FDA.

[11] Memo of W. B. White, March 9, 1933. Rec. Off. Sec. Ag.

[12] *ibid.*

[13] Notice to Growers and Shippers of Fruits and Vegetables, June 20, 1933. Rec. Off. Sec. Ag.

[14] For an example of these liberalization appeals, see the letter of R. E. Trumble, Wenatchee, to the FDA., March 20, 1936. Rec. Off. Sec. Ag.

[15] S. Marcovitch, et al., Tennessee AES, *Bulletin 162*, 1937, p. 38.

[16] Smith and Leverton.

[17] S. Marcovitch, *Amer. Pomol. Soc., Proc.*, *54*, 147-149 (1938); *ibid.* and W. Stanley, *J. Econ. Ent.*, *31*, 480-482 (1938).

[18] Letter from Campbell to Secretary of Agriculture, Jan. 11, 1938. Rec. Off. Sec. Ag.

[19] Letter from H. A. Wallace to F. Lillie, President, National Academy of Sciences, February 1, 1938. Rec. Off. Sec. Ag.

[20] Notice to Growers and Shippers, Nov. 14, 1938. Rec. FDA.

[21] Notice to Growers and Shippers of Fruits and Vegetables, February 21, 1933. Rec. FDA. The coincidental increase in lead residues is discussed in *Federal Food, Drug, and Cosmetic Law, Administrative Reports, 1907-1949*, Washington, 1951, p. 789.

[22] Tugwell's intervention is discussed by Cavers, *Law and Contemporary Problems, 6,* 15 (1939).

[23] Tugwell, "Recollections of '33 and Later," *FDA Papers,* June 1968, pp. 4-5.

[24] *ibid.,* p. 4. Wallace's belief that his job was "to protect the farmers, not the consumers," is referred to by Tugwell again in "R. G. Tugwell, An Interview with Charles O. Jackson, June 7, 1968." Transcript in Oral History Collection, National Library of Medicine, Bethesda, Maryland.

[25] Notice to Growers and Shippers, June 20, 1933. Rec. FDA.

[26] These letters abound in Rec. FDA.

[27] Memorandum attached to letter from H. A. Wallace to F. Lillie, November 10, 1936. Rec. Off. Sec. Ag.

[28] *ibid.,* June 27, 1936.

[29] These articles are published in *J. Pharmacol. Exp. Therap., 64,* 364-445 (1938).

[30] Letter from R. G. Phillips to W. G. Campbell, July 16, 1931. Rec. Off. Sec. Ag.

[31] Report of Conference Concerning Spray Residue Regulations, p. 7. Rec. FDA.

[32] W. A. Ruth, *Amer. Pomol. Soc., Proc., 51,* 137-149 (1935); *ibid., 57,* 160-173 (1941).

[33] For an evaluation of Cannon's congressional career, see *Memorial Services Held in the House of Representatives and Senate of the United States, Together with Remarks Presented in Eulogy of Clarence Andrew Cannon, Late a Representative from Missouri,* Washington, 1964.

[34] Letter from Cannon to H. L. Brown, Department of Agriculture, Sept. 12, 1938. Rec. Off. Sec. Ag.

[35] The Cannon episode is discussed by F. B. Linton, *Food, Drug, Cosmetic Law J., 5,* 483 (1950).

[36] *ibid.,* pp. 484-485.

[37] *U.S. Statutes at Large, 50,* 396 (1937).

[38] House of Representatives, *Hearings Before the Subcommittee of the Committee on Appropriations, Treasury Department Appropriation Bill for 1938,* p. 875.

[39] *U.S. Statutes at Large, 52,* 135 (1938).

[40] I. Sturman, *Health and Hygiene,* November 1937, p. 139.

[41] F. J. Schlink, *Con. Res. Bull., 4* (n.s., 6), 23 (1938).

[42] L. T. Fairhall and P. A. Neal, *Pub. Health Reports*, *53* (ii), 1,245 (1938).

[43] A. J. Carlson, *J. Am. Dietetics Assoc.*, *15*, 1-5 (1939).

[44] Letter from Cannon to Department of Agriculture, Sept. 8, 1938. Rec. Off. Sec. Ag.

[45] Letter from Cannon to H. L. Brown, Sept. 12, 1938. Rec. Off. Sec. Ag.

[46] Notice to Growers and Shippers from H. A. Wallace, Sept. 19, 1938. Rec. Off. Sec. Ag.

[47] *Big "Y" Bulletin*, Sept. 1938.

[48] Letter to P. Appleby, Undersecretary of Agriculture, Nov. 13, 1940. Rec. Off. Sec. Ag.

[49] Letter to W. Campbell, Dec. 6, 1938. Rec. FDA.

[50] Copy enclosed with letter to Roosevelt, Sept. 27, 1939. Rec. Off. Sec. Ag.

[51] P. Dunbar, *Food, Drug, Cosmetic Law J.*, *14*, 135 (1959).

[52] Quoted by Jackson, p. 27.

[53] *Business Week*, October 1933, p. 13.

[54] *ibid.*

[55] Lord, p. 346.

[56] Quoted by Dunbar, p. 136.

[57] Schlink, *Ann. Amer. Acad.*, *173*, 125-143 (1934); also see Cavers, pp. 9-10; and O. G. Villard, *The Nation*, *138*, 433 (1934).

[58] Cavers, p. 10.

[59] *Big "Y" Bulletin*, Dec. 1933; June 1938.

[60] Lamb, p. 251.

[61] Jackson, p. 179f., offers a detailed discussion of the political intrigues associated with the tolerance provision of the Copeland bill.

[62] Cavers, p. 21.

[63] *Congressional Record*, *83*, 7,779 (1938).

[64] Quoted by Jackson, p. 181.

[65] *ibid.*, p. 186.

[66] See, for example, *Big "Y" Bulletin*, June 1938.

[67] The following statistical information has been gleaned from the FDA annual reports, published in *Federal Food, Drug and Cosmetic Law, Administrative Reports, 1907-1949*, Washington, 1951.

[68] *ibid.*, p. 931.

279

[69] *ibid.*, p. 853.

[70] *ibid.*, p. 876.

[71] P. A. Neal, et al., *U.S. Pub. Health Bull.*, *267*, 1941, p. 86.

[72] Carlson, p. 3.

[73] *Federal Food, Drug, and Cosmetic Law*, p. 990.

[74] *Consumers' Union Reports*, *5*, Sept. 1940, p. 18.

[75] T. J. Talbert, *Amer. Pomol. Soc., Proc.*, *56*, 126 (1941).

[76] Letter from H. A. Wallace to Sen. H. T. Bone, April 12, 1940. Rec. Off. Sec. Ag.

[77] Dunbar, *Food, Drug, Cosmetic Law J.*, *4*, 234 (1949).

[78] *Federal Register*, *19*, 6,738-6,772 (1954); for tolerance hearings, see Hearings held before the Administrator, Federal Security Agency, in the matter of: Tolerances for Poisonous or Deleterious Residues on or in Fresh Fruits or Vegetables, transcript docket FDC-57, HEW accession no. 68A-352, cartons 15-18. NRC.

Epilogue

[1] *Reader's Digest*, November 1945, p. 84.

[2] There is much material available on the early history of DDT. The chief source used for this discussion was J. Leary, W. Fishbein and L. Salter, *DDT and the Insect Problem*, New York, 1946.

[3] D. Stick, *Pop. Sci. Mon.*, *146*, 155 (1945).

[4] O. T. Zimmerman and I. Levine, *DDT: Killer of Killers*, Dover, New Hampshire, 1946; *Am. J. Pub. Health*, *36*, 657 (1946).

[5] *Collier's*, *116*, 27 (1945).

[6] *Reader's Digest*, p. 84.

[7] V. Wigglesworth, *Brit. Med. J.*, 1945 (i), p. 517.

[8] G. Woodward, *Science*, *102*, 177-178 (1945); O. G. Fitzhugh and A. A. Nelson, *J. Pharmacol. Exp. Therap.*, *89*, 18-30 (1947); P. A. Neal, et al., *Pub. Health Rep., Supp. 177* (1944); *ibid., Supp. 183*, 1945.

[9] J. K. Terres, *The New Republic*, *114*, 415 (1946).

[10] *Silent Spring*, New York, 1962, p. 157. Miss Carson's life and work have been examined by P. Brooks, *The House of Life*, New York, 1972.

[11] E. Diamond, *Sat. Eve. Post*, *236*, 16 (1963).

[12] *Silent Spring*, p. 22.

Index

Académie de Médicine, 89-91
Accum, Frederic, 97
acetanilid, 197n
Adams, Samuel Hopkins, 186
advertising, 186-187
"Affaire des Poisons," 44
agricultural experiment stations, 11
Agricultural Revolution, 3
agriculture: effect of transportation, 4; machinery, 5; post-World War One depression, 129
alcohol, 197n
Alpher, Isadore, 194
Alsberg, Carl L., 117, 118, 123, 126, 137, 138, 155
American Agriculturalist: on spread of potato beetle, 18; endorsement of Paris green, 21; on farmers' fears of Paris green, 29; assurance of safety of sprayed produce, 35
American Bee Journal, 27-28
American Chamber of Horrors, 204, 205, 206
American Entomologist, 12
American Journal of Public Health, 178

American Medical Association, 60, 179
American Pomological Society, 174, 227, 228
American Public Health Association, 179
Anderson, N. P., 179
aphid, 13
arsenic: use as insecticide, 17; definition, 20; effects on bees, 27-29; toxicology, 36-37; acute poisoning by, 38; chronic poisoning by, 38-39; presence in manufactures, 39-40; in wallpapers, 40-44, 59; history as therapeutic agent, 49f.; use as cosmetic, 50; criticism of therapeutic use, 51; as cause of cancer, 51f.; "world tolerance," 82, 87, 123, 137; use as insecticide in France, 88; as normal body constituent, 123, 177; contamination of environment by, 176-177; in seafood, 183; poisoning of soil by, 217
Association of Economic Entomologists, 28
Aub, Joseph, 158

281

Stevenson, Thomas, 85
Styrian arsenic-eaters, 53-57,
    78, 183
sulfurous acid and sulfites,
    103, 105, 109, 190
Summers, J. W., 123, 124
Suncrest Orchards, 141, 142
Supreme Court, 197

Taft, William Howard, 197
Talbert, 182-184, 246
Tarbell, Ida, 185
tartar emetic, 47
Tennessee Agricultural
    Experiment Station, 217, 219
therapeutic skepticism, 47
Throne, Binford, 178, 179
tobacco, 15, 207-208
Tofana, 44
Tonney, Frederick, 147
*Treatise on Adulterations of
    Foods and Culinary Poisons, A,*
    97
tri-sodium phosphate, 220
Trouvelot, Leopold, 23
Tugwell, Rexford, 199, 200,
    221-222, 236, 237
Tusser, Thomas, 15
typhus, 248

Uncle Tom's Cabin, 101n, 187
urbanization, 4

vedalia beetle, 13
Vlcek, Fred, 79-80, 145
Voegtlin, Carl, 155
Vogel, Karl, 176
von Tschudi, Johann, 54, 55, 56

Wallace, Henry A., 219, 222,
    225, 241, 242
Wallace, Henry C., 130
Walsh, Benjamin D., 8, 9, 12
Washington, George, 3-4
Washington State Grange, 234
Washington State Horticultural
    Association, 208

Waterman, W. C., 173-174
Webster, F. M., 28-29
Wedderburn, Alexander, 100
Weiss, M., 89-90
Wharton, W.R.N., 131
Wiley, 183n, 235; apiculturists'
    hatred of, 28n; background,
    99; campaign against
    adulteration, 99-101;
    opposition to synthetic food
    ingredients, 103-104; poison
    squad studies, 104-105;
    hearing with Theodore
    Roosevelt, 106; conflict with
    Remsen Board, 107; views on
    chronic poisoning, 107-108;
    opposition to food additives
    for cosmetics purposes,
    109-110; action against
    sulferous acid treated fruit,
    114; expectation of industrial
    cooperation with food
    regulation, 115-116; resignation
    from Bureau of Chemistry,
    116; employment by *Good
    Housekeeping,* 117; desire for
    stricter enforcement of food
    law, 117; *Foods and Their
    Adulteration,* 120; mention of
    spray residues, 120-121;
    criticism of Waterman
    Amendment, 174; criticism of
    patent medicine regulation,
    197
Wilson, James, 115, 117
Winslow, John, 51
Woefel, 79, 80
Woods, C. D., 76, 87
Woodward, William C., 95, 96,
    122, 123, 124, 137, 138

Yakima Chamber of Commerce,
    167
Yakima Fruit Growers'
    Association, 171
Yakima Valley, 73, 167, 170, 203
Your Money's Worth, 187-188

Library of Congress Cataloging in Publication Data

Whorton, James C   1942-
  Before Silent Spring.

    1. Pesticides—Toxicology. 2. Pesticide
residues. 3. Food contamination. 4. Food
adulteration and inspection—United States—History.

I. Title.
RA1270.P4W45   632'.95'042   74-11071
ISBN 0-691-08139-5

9 780691 618296